EN–
TAN–
GLE

PHYSICS AND THE ARTISTIC IMAGINATION

EN–
TAN–
GLE

PHYSICS AND THE ARTISTIC IMAGINATION

BILDMUSEET
UMEÅ
UNIVERSITY

HATJE
CANTZ

CONTENTS

NEXT SPREAD:
Detail
Remains
Davide Quayola
2018

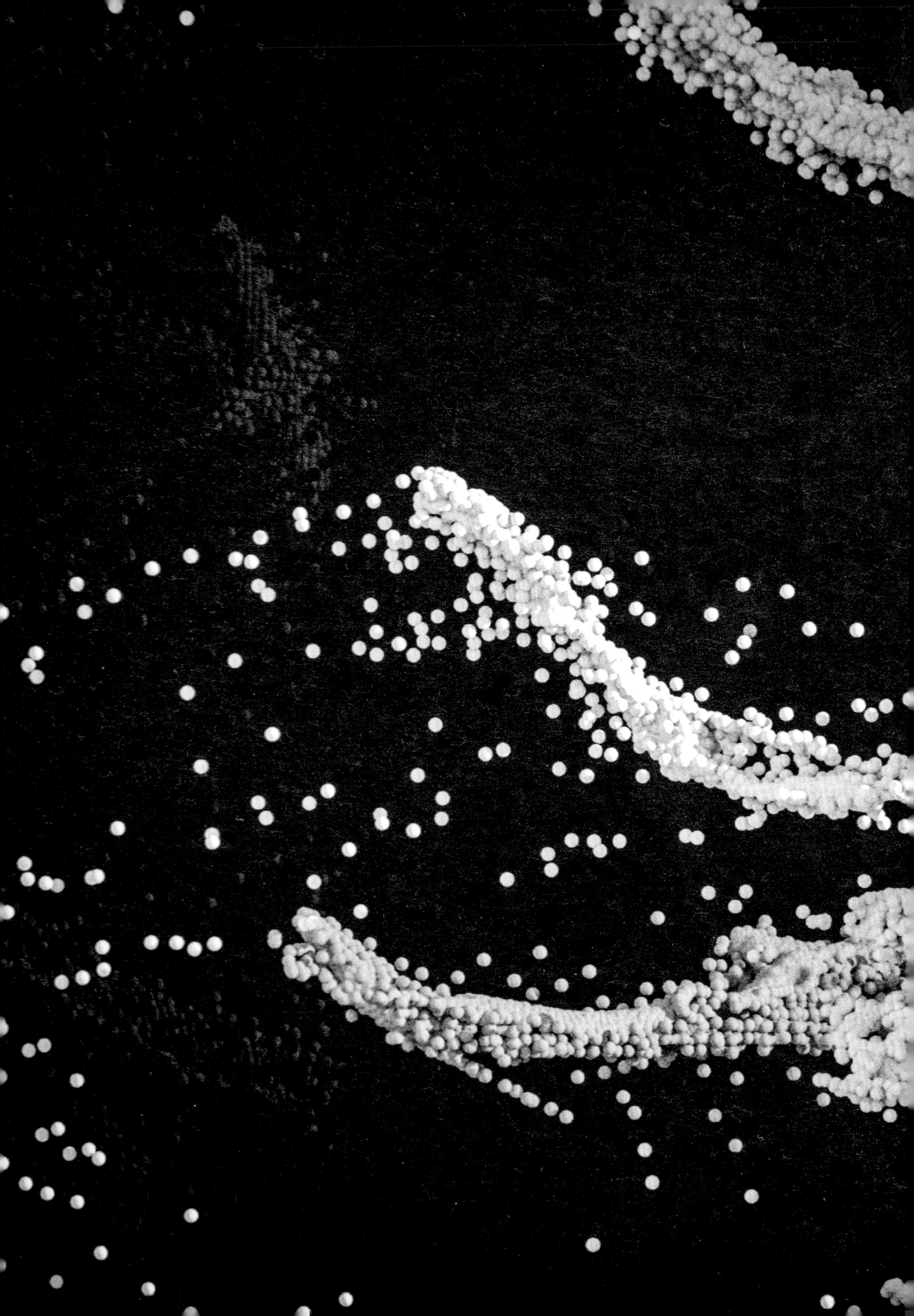

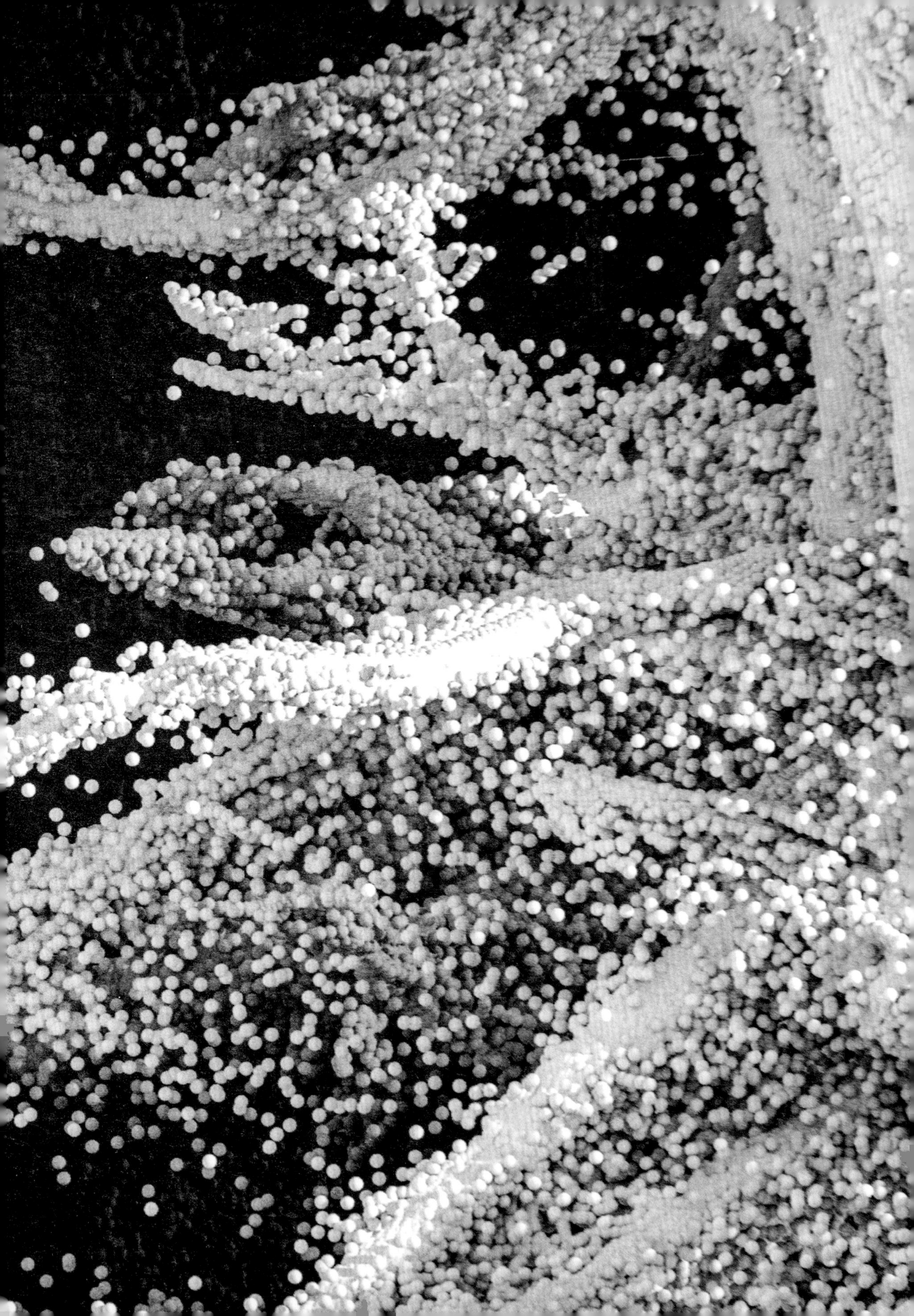

PREFACE

Black holes, dark matter, gravity, space, time, and quantum entanglement are phenomena that fascinate scientists and artists alike. The exhibition *Entangle: Physics and the Artistic Imagination* and this accompanying publication present painting, installation, fashion, architecture, sculpture, film, digital artwork, and photography by fourteen internationally recognized artists who are inspired by the science that reveals the fundamental forces shaping our world. Together with the thought-provoking artworks, this publication also presents interviews with artists and physicists who share their different ways of seeing. In addition, key scholars in the field have generously contributed their essays on the critical encounter between art and physics.

The curator of *Entangle: Physics and the Artistic Imagination* is Ariane Koek, founder of the Arts at CERN program at the European Organization for Nuclear Research in Switzerland – a program which invites artists to create work linked to the search for the universe's smallest component parts. A number of the artists in the *Entangle* exhibition and publication have participated in the CERN program, and some of the artworks were created especially for the exhibition at Bildmuseet.

Entangle: Physics and the Artistic Imagination is produced by Bildmuseet at Umeå University in Sweden. Bildmuseet is a public space for engaging with ideas in contemporary art together with, and alongside, current scientific and academic research – a creative space where different forms of knowledge and expertise can meet. As director of Bildmuseet, I am in a privileged position; I cannot imagine a more interesting and rewarding job. Together with my amazing team, I get to work with artists and curators to facilitate the realization of their new projects and articulate questions of urgency in contemporary society. For me, one of the shared qualities between art and science is that they challenge our preconceived ideas about the world, our ways of understanding the world and ourselves in it. Art and science propose another way of seeing and another way of being in the world – they force us to imagine the world differently.

Entangle is the result of a successful interaction. I would like to express my sincere and heartfelt thanks to all of the artists participating in the exhibition and the publication: Julius von Bismarck, Julian Charrière, Sou Fujimoto, Iris van Herpen, Ryoji Ikeda, William Kentridge, Rafael Lozano-Hemmer, Goshka Macuga, Davide Quayola, Solveig Settemsdal, Sarah Sze, Keith Tyson, Jorinde Voigt, and Carey Young. And to all of the physicists and essayists who have contributed to this publication: Philip Ball, Bilge Demirköz, Michael Doser, Peter Galison, Gavin Parkinson, Carlo Rovelli, Nicola Triscott, Subodh Patil, Kate Shaw, and James Wells.

A special thank you is extended to the private owners and institutions who have generously lent their works to the exhibition. My special word of gratitude goes to Ariane Koek, for her vision and commitment toward this remarkable venture and whose knowledge was a precondition for its realization. Finally, I would like to take the opportunity to thank my colleagues, the professional team at Bildmuseet.

KATARINA PIERRE
Director, Bildmuseet, Umeå University

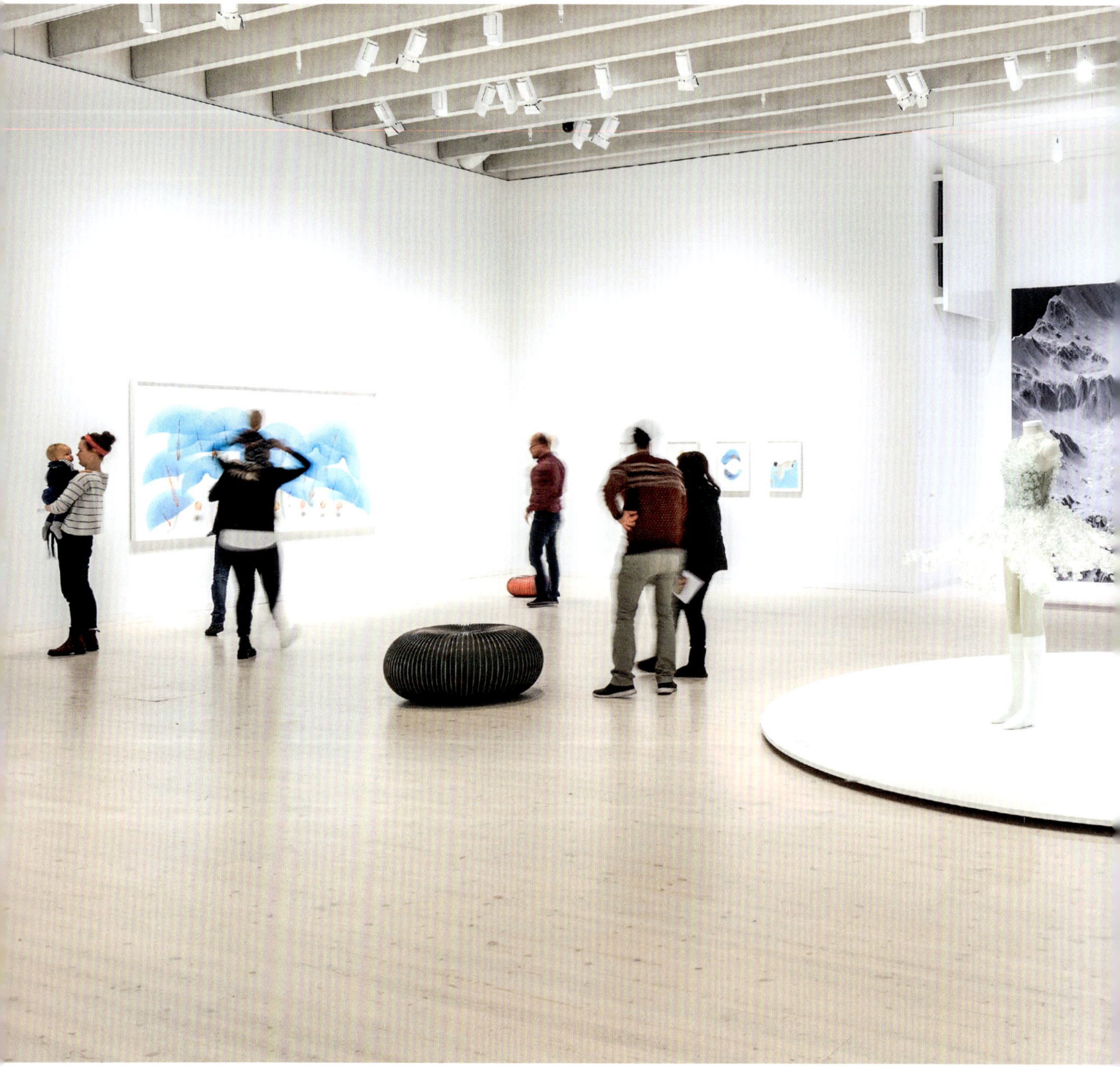

Installation view
Entangle: Physics and the Artistic Imagination
Bildmuseet, November 16, 2018 - April 7, 2019

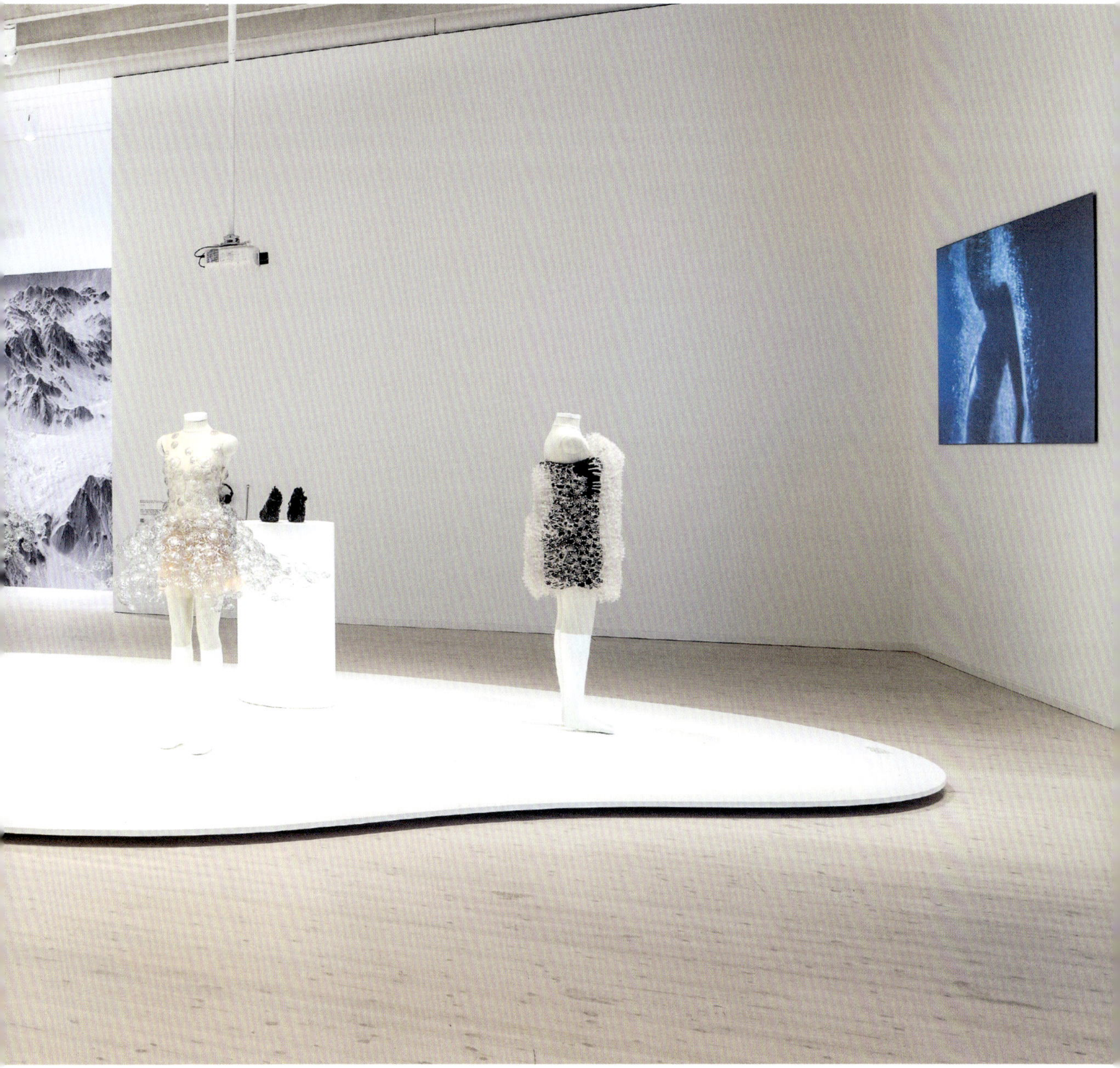

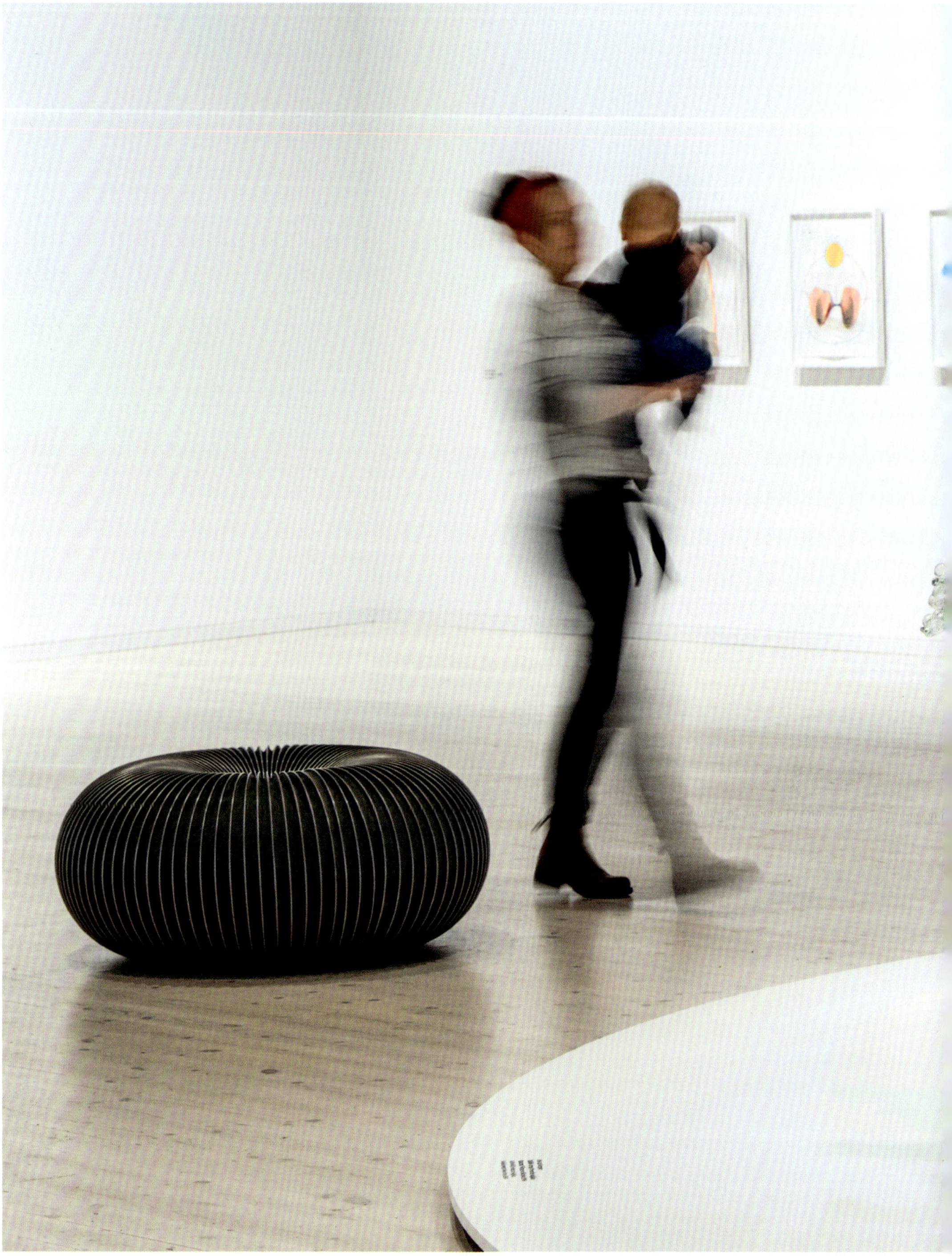

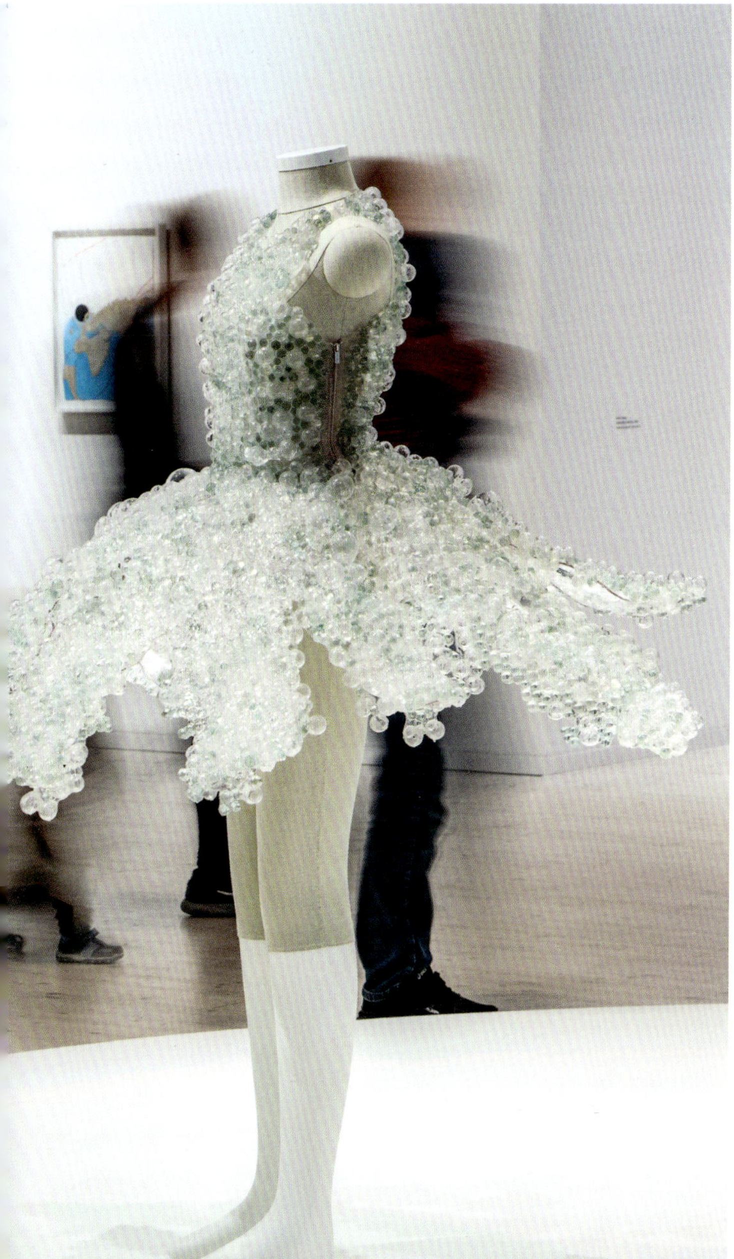

LEFT: *Two Self-Revolving Toruses*
(part of work)
Julius von Bismarck
2018

RIGHT: Bubble dress from the *Sejiaku*
Autumn/Winter collection 2016
Iris van Herpen

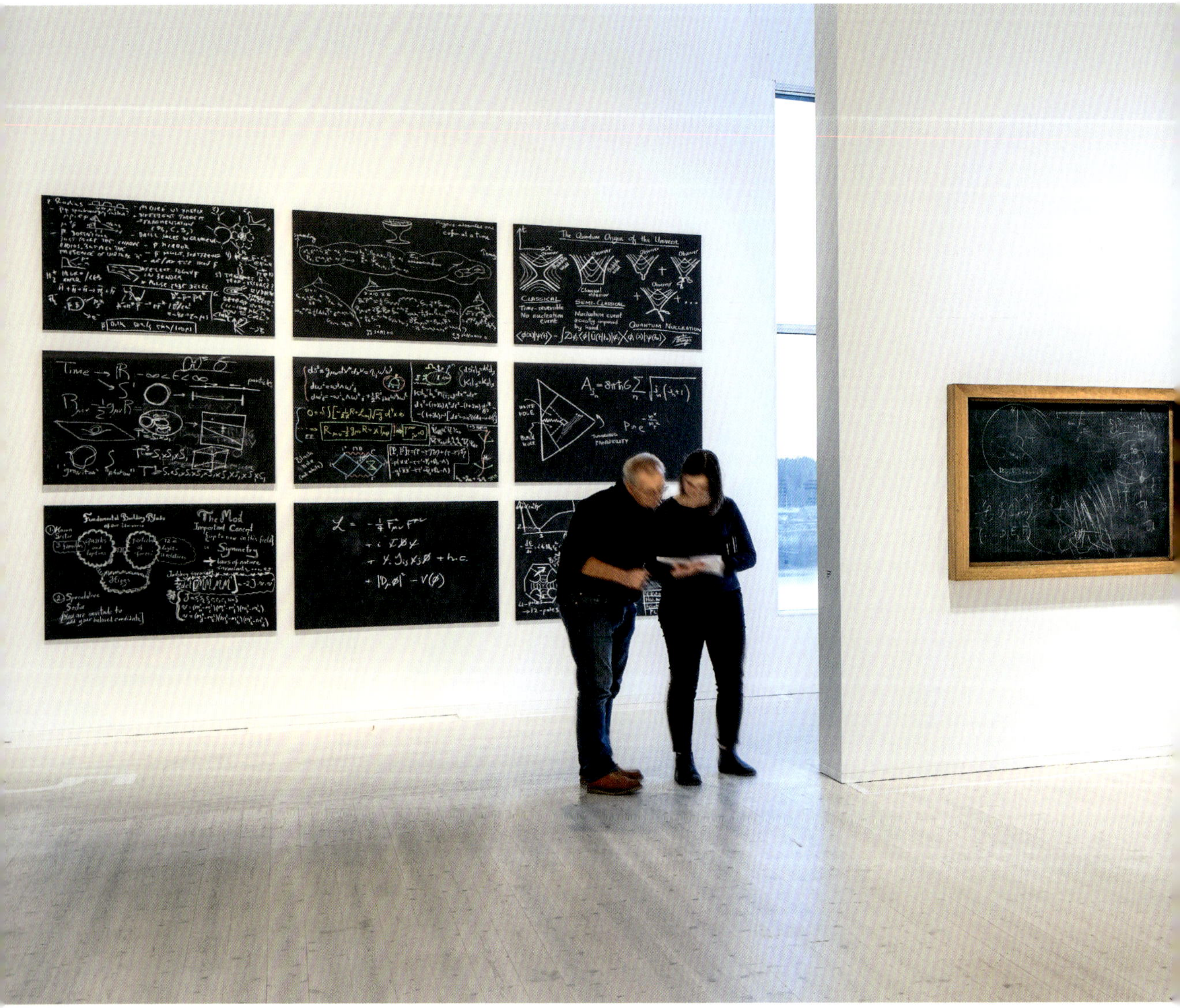

Untitled, installation by Goshka Macuga, 2018

Particle physics equations by contemporary physicists, 2018
(see pp. 76, 148–149)
Blackboard with white chalk traces, Albert Einstein, 1931
Continuum, Joseph Beuys, 1978/1984
Various drawings by Rudolf Steiner, 1921–1924

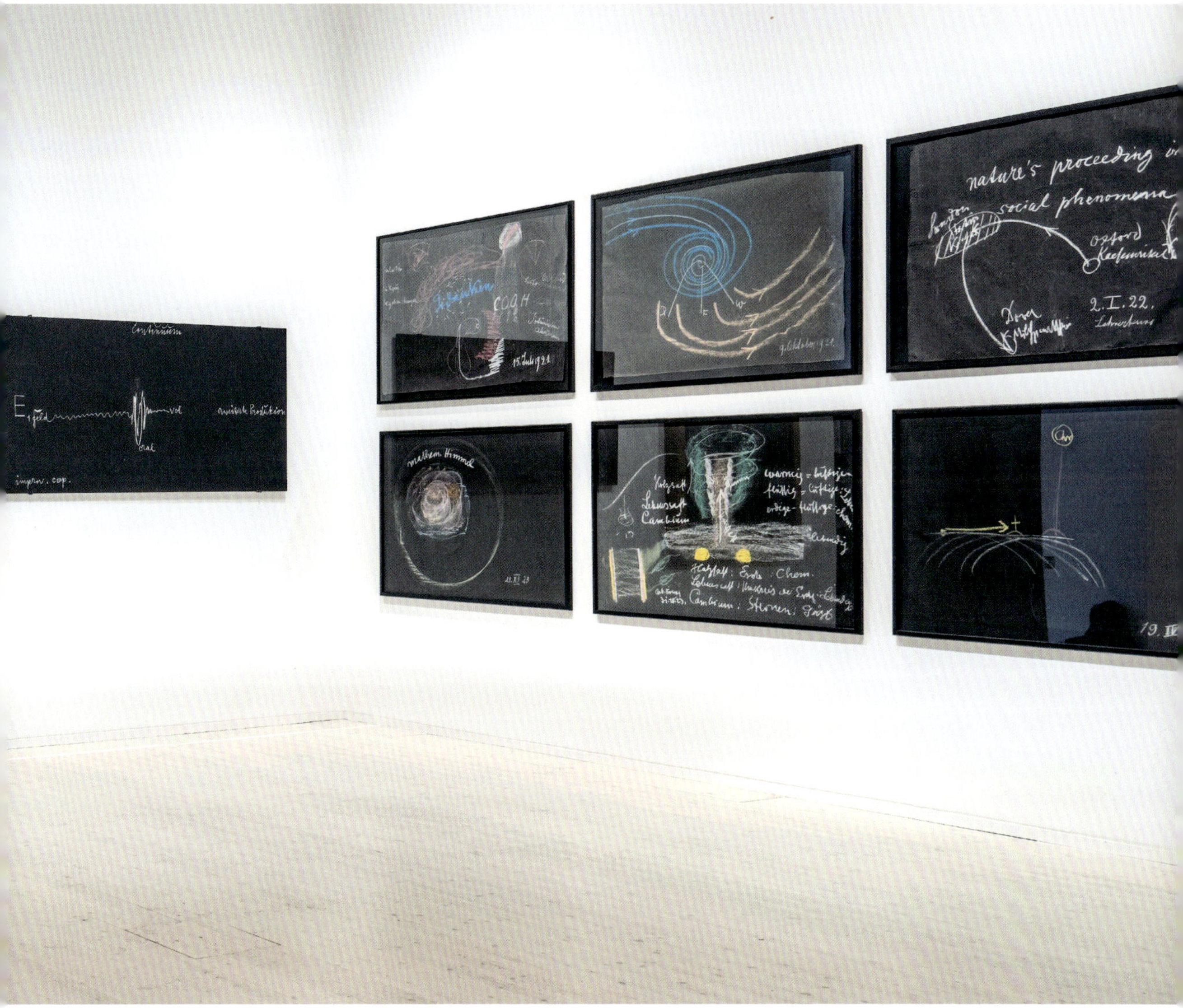
nature's proceeding
social phenomena
Oxford
2. I. 22.
Cambium

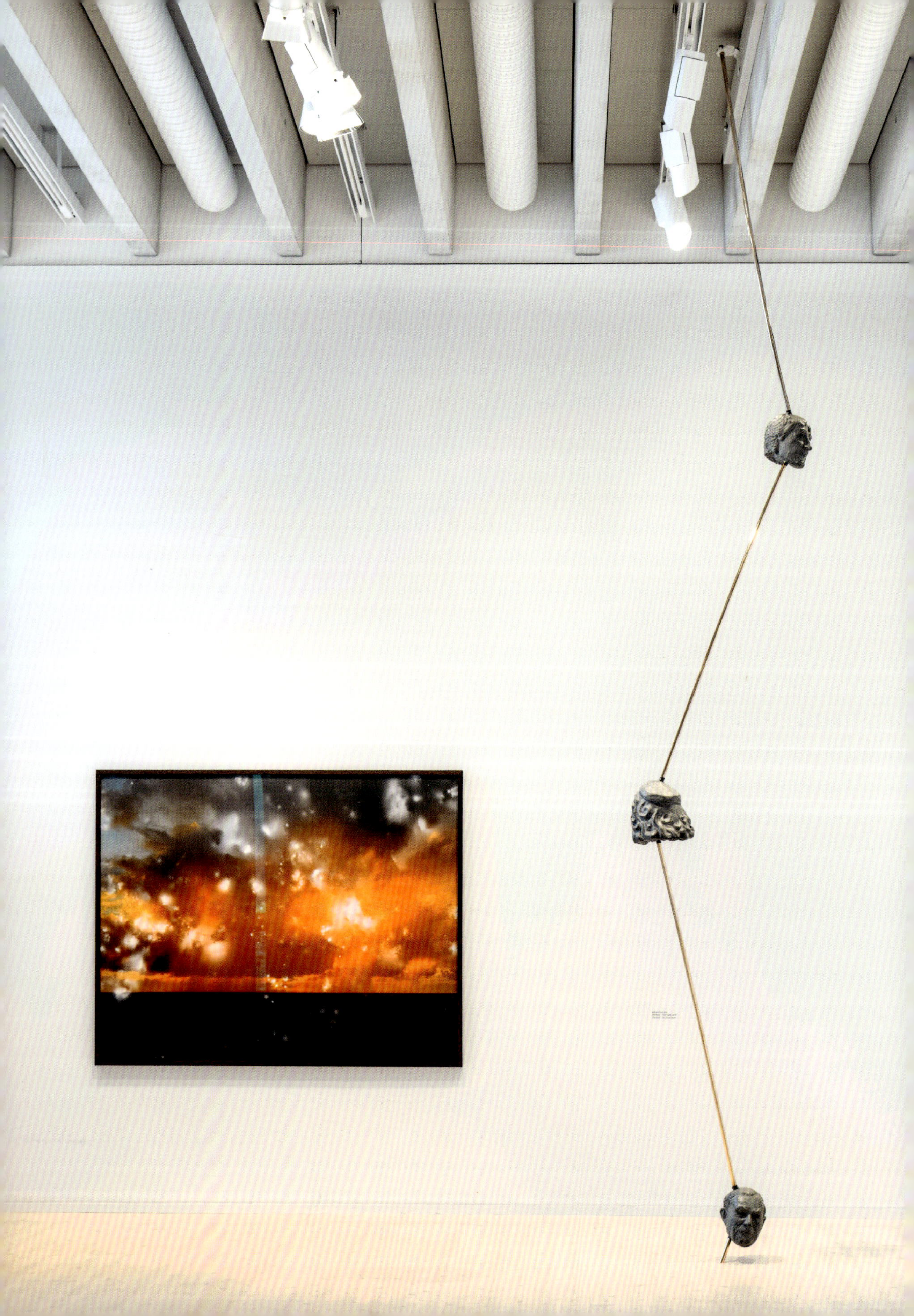

INTRODUCTION TO ENTANGLE: PHYSICS AND THE ARTISTIC IMAGINATION

Ariane Koek

> The very nature of materiality is an entanglement. Matter itself is always already open to, or entangled with, the "Other." The intra-actively emergent "parts" of phenomena are co-constituted. Not only subjects but also objects are permeated through and through with their entangled kin; the other is not just in one's skin, but in one's bones, in one's belly, in one's heart, in one's nucleus, in one's past and future.
>
> KAREN BARAD

> Everything we call real is made of things that cannot be regarded as real.
>
> NIELS BOHR

We are all made of star dust. Matter is 95 percent empty. Our feet are older than our heads. The ground beneath our feet is full of holes. We are made of stuff which is older than the earth we inhabit.[1]

Hearing these kinds of statements, the material reality we live in is blasted apart. These everyday observations by particle physicists show how physics can open up new realms of perceiving reality which take us into different dimensions. Our material world takes on a whole different existence, and the things we cannot see (and do not regard as real) create, as Niels Bohr indicates, our reality. The invisible makes the visible, the immaterial, matter, and vice versa. Separation is illusory. Everything is entangled in a state of intra-connection and constant becoming.

From the tiny to the vast, technology is now allowing us to observe matter in the micro to the macro in detail we have never been able to see before: from a supernova exploding in a galaxy about 930 million light years away across the vastness of the universe to the measurement of gravitational waves below the level of the nucleus with astounding accuracy. New material realities are constantly opening up in scales and details which challenge our perceptions. We are in an era of new phenomenologies.

LEFT: *Flathead - First Light*
Julian Charrière
2016

RIGHT: *International Institute of Intellectual Cooperation, Configuration 7 (continued)*
Goshka Macuga
2015-2016

Entangle: Physics and the Artistic Imagination is dedicated to works by contemporary artists working in art and design in the twenty-first century who are inspired by particle physics. This is the science which investigates the fundamental constituents of matter – particles – as well as the invisible structures and the forces of nature which shape and govern our world. Artists are usually credited with allowing us to see the world through a different lens. Physics offers artists an additional lens for looking at the world differently, thus providing them with another compelling source of inspiration for their imaginations and their creative practice.

But physics and the arts share a commonality. Both physics and the arts explore the fundamental questions about how we came into existence and try to make sense of our place in the world. Both set out to discover answers to these questions. However, they express the process and answers in different ways: physics through the abstract language of mathematics, which is then brought down to earth and tested out by scientific methodology, concrete engineering, and experimentation; the arts through the senses and making objects and/or experiences which reflect these discoveries and its process. Artists and particle physicists also share a very particular affinity. Particle physicists are divided into theorists who think in the abstract, beyond the paradigms, and the experimentalists who test these theories. In a sense, an artist is a combination of both: s/he is both a theorist and an experimentalist even though the ways of knowing and expressing knowledge are so different.[2]

What is entangled in the process of both of these different ways of knowing and looking at the world is crucially the imagination – that mysterious process by which we make unexpected links, out-of-the-box connections, and have inexplicable intuitions beyond the known world. This exhibition deliberately takes as its focus artists who for part of their practice use physics as springboards of their imaginations.

This is just one particular focus within what is often referred to as the arts/science field. Since 1967 this field has been expanding with ever-increasing energy: with the founding of Experiments in Art and Technology (E.A.T.) at the Bell Laboratories initiative, headed by the engineer Billy Kluver and the artist Robert Rauschenberg, which spearheaded the co-creation of new work across the disciplines, and the pioneering work of the Hungarian émigré artist György Kepes, who founded the Center for Advanced Visual Studies at the Massachusetts Institute of Technology (MIT), Boston. Today, in the twenty-first century, it is particularly popular with new initiatives being announced every week. This popularity is due to a number of factors: economics – the reduction of funding for the arts and humanities, with science being “relatively” better funded; the rise of public engagement as a priority for science organizations, including reaching beyond their traditional audiences; and, in a world beset with global challenges, the drive to innovate which is understood to happen when disciplines mix.[3] Add to this also the eternal hunger of artists to find new sources of inspiration, as well as a world in which science, not art, is now seen as the ultimate source of wonder, and the key factors for art science’s ascendancy are very clear.[4]

However, the term arts/science or even sci art, which is applied to this field, suggests that it is one unified and homogenous movement. It isn’t. In reality it is a multiplicity of movements which go in many different directions, and which are defined and driven by many different intentions, seeking different outcomes and having very different and varying aesthetics.[5] For example the arts/science work of Ars Electronica, ZKM, and Music Tech Fest can be defined as being driven by the exploration of the limits and possibilities of technology. That of the Science Gallery Dublin and the Exploratorium in San Francisco is defined by its intention to engage the audience, with its emphasis on public engagement and interaction with the ideas of science aided by artists and designers. Sometimes these movements overlap because the intentions are mixed, sometimes they don’t. Sometimes the emphasis is on the scientific side, sometimes it is more on the artistic, or there is a balance in between.

There are artists who use the tools, experiments, and techniques of science to create their artwork, as in the case of Evelina Domnitch and Dmitry Gelfand, who merge physics, chemistry, and computer science, and of the British artist Anna Dumitriu, who works with digital and biological media. The work of the painter Keith Tyson, who features in *Entangle,* deliberately comes from the perspective of science being imaginative fuel for artists who he calls, together with scientists, “knowledge seekers.”[6] This approach taken by the exhibition is not new, but it is significant both today and in the past. As the art historian Gavin Parkinson points out in his essay in this catalogue, there is a long tradition of science as fuel for the imagination of artists. For example, the Surrealist manifesto by André Breton was highly influenced by physics; and physics remained an important source for Salvador Dalí until he died.

Because of the multiplicity of movement with its diversity of intentions and outcomes, which are potentially entangled, there's an argument that the term art science or sci art is redundant and meaningless – or at the very least should be treated with caution.[7] What is perhaps far more important is the notion of transdisciplinarity – that intellectual and creative space where different forms of knowledge and expertise meet across, around, and on their boundaries and borders. I founded the Arts at CERN program on this premise. So much so that the focus in the program in its first five years was very much more on the process and the encounters between the artists and the physicists in the laboratory in Geneva, rather than on any output.[8]

The thinking was to get the curation of the process of the interactions between artists and scientists right, and then the creative work will naturally happen without any deadlines whilst the minds and hearts of the artists and physicists collide to spark the imagination. Four of the artists in the exhibition were part of the first years of the Arts at CERN program. Julius von Bismarck, a former student at Olafur Eliasson's famous interdisciplinary school Institut Für Raumexperimente, Berlin, was the first Collide International artist-in-residence, spending two months in total at the laboratory. Physics inspires his daily creative practice to this day. He was followed two years later, in 2014, by the Japanese musician and digital artist Ryoji Ikeda, the third Collide International artist who made two large-scale installations as a result of his CERN encounter. The other two artists, the Dutch fashion designer Iris van Herpen and the Polish artist Goshka Macuga, participated in the Visiting Artists program, with its one day of intense visits and briefings on the latest findings in physics. Both artists continue to visit CERN in order to refresh their knowledge and blast their imaginations into new dimensions.

Space, Time, Light, Gravity, Matter, Entropy – such themes preoccupy the fourteen artists in *Entangle*. And to make this clear at Bildmuseet, a contemporary art gallery which eschews explanatory texts on its walls, there was an accompanying audio-cloud at the heart of the exhibition. This was experienced online, as a podcast, in the exhibition on headsets or mobiles, and in the catalogue. In "diptychs" artists and physicists talk about the same phenomena from their different standpoints. Recorded at different times and in different places, it is only on the page and in the air that they and their reflections on different phenomena "meet" for the first time.

For example, the Turkish physicist Bilge Demirköz, who has helped build the Alpha Magnetic Spectrometer (AMS) detector on the International Space Station, reflects on the phenomenon of space and what it means to her. Her words are alongside those of the architect Sou Fujimoto, who is known for turning architectural concepts of space inside out and who confesses that for him outer space is terrifying because of what he calls its lack of density. For him, trees are the ultimate symbols of space.

In her interview, the fashion designer Iris van Herpen reveals how matter is her inspiration as well as her greatest enemy in the process of creating new materials and forms. Her thoughts are laid alongside those of one of the world's leading experts on antimatter, Dr. Michael Doser, who reveals the duel to the death of matter against antimatter, the innate combative nature of particles that led to the creation of the universe and matter itself.

These juxtapositions reveal the differences as well as connections of seeing the same phenomena through different eyes. It is in the connections, the differences, and the gaps in between that the imagination thrives and creativity grows. We are forced to think and imagine beyond our own parameters, particularly by not knowing and by our failure to fully understand. After all, we only learn to walk as a child by falling down and picking ourselves up, time and time again, until we know our own limits and once we have tested gravity, like amateur physicists, to our utmost capability. Knowledge is built on the unknown and failure, with imagination daring to take us beyond our physical and conceptual limits, as Julius von Bismarck's video piece *Escape Shapes* (2016) in the exhibition indicates.

Julius himself says: "What makes it [physics] so inspiring for me is that you can always discover something that you couldn't understand or even imagine before and that you maybe still can't imagine. This boundary of my imagination is where I get a lot of inspiration. Every new weird fact that is funny to imagine or hard to imagine is for me an inspiration to go into a corner of my brain that I haven't used before or that is not existing yet. I have to kind of extend my imagination to understand it. Then, I'm failing in that. I like that failure. That's actually an inspiration for a couple of works I made."

Or as the physicist Albert Einstein famously said in a renowned newspaper interview published in *The Saturday Evening Post* in 1929: "Knowledge is limited. Imagination embraces the whole world."[9] This statement is dissected in Philip Ball's essay in the catalogue, in which he shows that the imagination is not just the exclusive domain of artists: it is fundamental to both experimental and theoretical physicists and the actual process of discovery.

The role of the imagination in discovery was also considered when the best-selling physicist Carlo Rovelli, who has written the lead essay for the catalogue, was drawing on one of the blackboards that form part of the Polish artist Goshka Macuga's new commission for Bildmuseet. It is a composite installation comprising blackboards drawn on by nine contemporary physicists, the German artist Joseph Beuys, Albert Einstein, and the founder of biodynamics and educationalist Rudolf Steiner. When Rovelli wrote an equation using the Penrose system for describing his theory of the white hole, the theoretical opposite counterpart to the black hole, he said that just because white holes in the universe haven't been seen yet doesn't mean that they don't exist. "You just have to imagine them." He added that when black holes were first predicted by Einstein in 1916 with his general theory of relativity, people said they did not exist. Fifty-five years later, the first one was observed and now they are commonly accepted as real.

Thus, the notion of science as being purely objective and logical is a myth. Imagination and intuition are entangled in the process of scientific discovery, just as they are in the artistic process. Our perception of science as the ultimate objective truth, which precludes the imagination, dates from the Enlightenment – the so-called Age of Reason, in the seventeenth century. This false separation may be due to a fear of science's origins when it was closely associated with alchemy – which is often characterized as the search for the philosopher's stone – the key to the material world and the transformation of matter, including base metals to gold. Science, in particular chemistry and the natural sciences, became detached from the perceived magic of alchemy by creating scientific methodology, factual proof, and analysis. But the imagination is very closely connected to magic. As Philip Ball points out in his catalogue essay, imagination shares the same etymological root as "magic" – "which, in the age just before the time of Isaac Newton, did not necessarily mean superstitious agency but the 'hidden forces' by which natural magicians comprehended and claimed to manipulate nature."

However, there may be a more complex reason for the separation of the imagination from science. It may relate to power and social control. The Italian philosopher Federico Campagna in his book *Technic and Magic: The Reconstruction of Reality* starts on the premise that reality is always a reconstruction.[10] Reality is never real per se. In fact, this is certainly what science says. We don't ever actually see an object in the real world: we only see the light bouncing off it, which is reflected in our eyes, and then our brains create an impression of the object which we then "see." He charts from the Age of Enlightenment the rise of Techne – logic, form, and order with its language of absolutism – as a way of controlling human beings and thus their behavior. Magic or the imagination poses a threat to Techne's social and political control of the masses because it centers its reconstruction of the world around the notion of the "ineffable," the unnamable and inexplicable which cannot be put into words or ever be pinned down, yet lies at the very heart of existence. This idea is not far away from the writings of the Romantic poets, such as Percy Bysshe Shelley, who wrote about the radical ability of the imagination to challenge all borders and to create revolution. He also went so far as to claim that the imagination is the "great instrument of the moral good."

I would go so far as to also say that this is why today contemporary art is becoming persistently driven to become utilitarian and more useful. Art is increasingly in danger of becoming a direct tool to describe, illustrate, and/or co-create solutions for political and social issues – what I would call "art by design." There is increasing pressure for contemporary art to have a purpose, system, and logic, which bleeds it of its mystery and ineffability – and thus also dissolves its imaginative and experiential powers as well as aesthetics. These powers also can transform lives, change attitudes, ways of looking at and experiencing the world – but they are not directly applied or measurable. Couple this with the rise of science and technology as dominant cultural and economic forces, and there is also the danger that science and technology together may swallow contemporary art whole – and become the more dominant part – to the detriment of art and ultimately of society.

All of these different factors create an urgent need to investigate and revise the way in which we understand the relationship between science and the arts, their histories, and how they intra-relate. One important step in this direction is a new book called *The Experimental Imagination* by the academic Tita Chico.[11] She shows that the arts and sciences in fact borrowed from each other in order to argue their legitimacy as modes of knowledge-making in Britain during the eighteenth-century Enlightenment. Literature used the tropes of science to legitimize its own claims to objectivity, whilst science used metaphors from literature as a way of explaining the imagination of science, giving it status by laying claim to be communicating to the popular realm. As her work shows, it is too simplistic to suggest that the Enlightenment heralded the division between objectivity and rationalism in science, and ego and experience in the humanities. The intra-connectiveness is far more complex than that. And we are only just beginning to unravel this entanglement.

Furthermore, Linda Dalrymple Henderson, the primary academic in the field of physics and art, based at the University of Texas at Austin, sheds new light on the relationship with the arts from the scientific point of view. She points out that ideas about space and time in fiction predated those of Einstein, which led to the general theory of relativity. Works of fiction may have opened up scientists' minds to exploring these possibilities in the first place.[12] This hints at what the eminent biologist Edward O. Wilson calls the extra dimension of the humanities, which science doesn't have: "the infinity of fantasy worlds."[13]

What is increasingly emerging in the fashion for art science or sci art is growing anecdotal evidence of how in fact the arts are influencing the work of the scientists themselves. For example, the theoretical physicist James Wells, who amongst other areas, looks at extra dimensions and new worlds, participated in the Arts at CERN program. He talked about how the experience of working with Julius von Bismarck made him look at his own theoretical work differently and would no doubt affect his findings in the future. The experimentalist and antimatter expert Michael Doser, one of the founding members of the CERN Cultural Board for the Arts program, similarly talked about how working with artists taught him how the artists' enjoyment of detours in making work caused him to look at alternative ways of problem solving his scientific experiments.[14] However, there is no proof that these interactions have led to big discoveries – yet. Much more academic investigation and extensive long-term tracking are required into how the arts have impacted upon science, and scholarship is only at the beginning. But what can be said without doubt, according to the former physicist and CEO/Artistic Director of Arts Catalyst, Nicola Triscott, is that the culture of physics as a whole is benefiting through what the philosopher John Dewey calls the "experiential" possibilities of art and its humanizing effect.

All of these factors point to the histories of arts and science being much more entangled than has been appreciated. There may very well be an "intra-action" between the arts and science – very much as there is in "tangential realism" – the approach created by the physicist turned philosopher Karen Barad. In her influential book *Meeting the Universe Halfway* from 2007, she talks about how everything is entangled – subject and object, inanimate and animate, sentient and non-sentient beings, stressing that

> To be entangled is not simply to be intertwined with another, as in the joining of separate entities, but to lack an independent, self-contained existence. Existence is not an individual affair.[15]

This is not simply inter-connection or inter-action. It is what Barad calls "intra-action," where the existence of one thing affects the existence and composition of another entity, whether sentient or not, across time and space. It's always changing, and always in the process of being. This is the entanglement described in the introductory essay by Carlo Rovelli, which also contains an honest and forthright confession: that we don't know *how* entanglement works the way it does. We just know *what* it does. As physics itself admits, we only know 6 percent of the universe. We have another 94 percent to learn more about, which apparently comprises the mysterious Dark Matter and Dark Energy which theoretically make up the rest of the universe.

For the next steps in scientific discovery, just as in art, what we need more of is greater curiosity, and above all our imagination, to drive knowledge and our different ways of knowing forward.[16] And perhaps that indicates what is really at the heart of all our existences: not only the intra-action between the visible and invisible, the material and immaterial, but also, crucially, "the ineffable." This cannot be contained/explained/reduced to a formula however hard we try in our rational and technological age. Like Sarah Sze's piece *Dark Matter,* seemingly made up of random detritus – Q-tips, photographs, postcards, glass tumblers, a reading lamp – what is entangled is that indefinable and illusive space on the borders of consciousness and unconsciousness, where imagination and knowledge meet. In that space, as in Sarah's work, we are in a constant state of entangled be-(com)ing – together (a)part.

[1] Anselm Kiefer, in his speech accepting the Heinrich Heine Prize in December 2014, spoke of the power of his visit to CERN under the Visiting Artists program earlier that year with the curator of his Royal Academy show, Kathleen Soriano. He recalls the scientists telling him that we are older than the earth because we carry with us particles that were present before the creation of our planet. See Anselm Kiefer, "Alexander, du bist ein Teilchenbeschleuniger," *Welt.de*, December 13, 2014, https://www.welt.de/kultur/kunst-und-architektur/article135326592/Alexander-du-bist-ein-Teilchenbeschleuniger.html (all URLs accessed in January 2019).

[2] Of all the sciences, particle physics is arguably the closest to the arts and artists, because it has all of these elements, including its intense philosophical questioning of what we see and what is behind it, as well as the nature of existence.

[3] *The Art Newspaper* online, October 4, 2011.

[4] Observation by the curator Kathleen Soriano: "Your point about the rise of science as a source of wonder made me think about how art was a source of wonder more when it was celebrating religion, or trying to inspire us to be more spiritual or religious. Now that there is no religion (hardly), is that another reason for science walking into that space?"

[5] In a lecture at The Exploratorium, San Francisco, October 2018, I outlined at least eight different forms of arts/science collaborations. I listed them as: artists who are driven by technology developed by or for science to create artistic work by exploring the limits and/or the aesthetics of technology; artists who use scientific practices, methods of experimentation and images to create new artworks; artists who work together with scientists to jointly create a new artwork together; artists who are combined with scientists to come up with solutions and products which address societal issues and global problems; artists who are used as the "canary in the coal mine" to be the voice of the people to enable scientists to look at their work from a more humanitarian and potentially ethical perspective; artists used to communicate and illustrate the science of the strand which is as old as the engagement of art and science itself; and the ideas of science as fuel and inspiration for the imagination of artists. In the future, it is not outside the realm of possibility that there may be an eighth strand – artists invited to help scientists to devise experiments in different ways. What defines all of these approaches is the point of departure and intention. The next day after the lecture, two of the attendees came up to me and said that they had listed at least another six.

[6] Ariane Koek, "Keith Tyson: the art of science and discovery," *CERN Courier*, February 24, 2010, https://cerncourier.com/keith-tyson-the-art-of-science-and-discovery/. "If you attempt to marry and equate art with science, then you fail. If you allow what is not similar about art and science, and their different methods and processes to co-exist and thrive, then a real art/science collaboration and aesthetic will emerge."

[7] Ken Arnold, "A Very Public Affair: Art Meets Science," *Interdisciplinary Science Review*, December 2017.

[8] Ariane Koek, "Creative Collisions - Beyond Paradigms," *Journal of the New Media Caucus*, 2013, http://median.newmediacaucus.org/isea2012-machine-wilderness/creative-collisions-beyond-paradigms-2/. Also see Ariane Koek, "In/visible: the inside story of the making of Arts at CERN," *Interdisciplinary Science Review*, December 2017.

[9] “What Life Means to Einstein: An Interview by George Sylvester Viereck,” *The Saturday Evening Post,* October 26, 1929, http://www.saturdayeveningpost.com/wp-content/uploads/satevepost/what_life_means_to_einstein.pdf.

[10] Federico Campagna, *Technic and Magic: The Reconstruction of Reality* (London: Bloomsbury, 2018).

[11] Tita Chico, *The Experimental Imagination: Literary Knowledge and Science in the British Enlightenment* (Stanford, CA: Stanford University Press, 2018).

[12] Peter L. Galison, Gerald Holton, and Silvan S. Schweber, eds., *Einstein for the 21st Century: His Legacy in Science, Art, and Modern Culture* (Princeton, NJ: Princeton University Press, 2008), p. xiii.

[13] Edward O. Wilson, *The Origins of Creativity* (London: Penguin, 2018), p. 187. “If science is the bedrock of the humanities, the humanities have further reach. Where scientific observation addresses all phenomena existing in the real world, scientific experimentation addresses all possible real worlds, and scientific theory addresses all conceivable worlds, the humanities encompass all three of these levels and one more, the infinity of all fantasy worlds.”

[14] Quotes taken from notes on conversations with the author in 2012–14.

[15] Karen Barad, in her influential book *Meeting the Universe Halfway: Quantum Physics and the Entanglement of Matter and Meaning* from 2007, published by Duke University Press.

[16] In 2011, the National Science Foundation in the USA, in collaboration with The Exploratorium, San Francisco, held a conference called Art as a Way of Knowing. This landmark conference with over 125 international artists, scientists, and transdisciplinary thinkers posited that the arts are a legitimate way of inquiry, research, and knowing as much as the sciences. See Marina McDougall, Bronwyn Bevan, and Robert Semper, “Art as a Way of Knowing Conference Report,” 2012, https://www.exploratorium.edu/files/pdf/cils/Art_as_a_Way_of_Knowing_report.pdf.

QUANTUM ENTANGLEMENT

Carlo Rovelli

There is a real but mysterious connection between things that are far apart. It has excited imagination and fantasy among scientists and nonscientists alike. It is called "entanglement."

It is perhaps the weirdest of the phenomena that physicists discovered in the last century, working around a weird theory of nature: quantum theory. It keeps puzzling scientists and everybody else who gets to know about it.

If you take a pair of gloves and ship one to New York and one to Shanghai, then the two gloves are widely separated, but they are in a sense related. In the following sense: if the one in New York is a left glove, then for sure the one in Shanghai is a right glove. If you are in New York, receive one of the boxes, and find a left glove in it, then you immediately know that the glove in Shanghai is a right glove. This connection between faraway things is not surprising, not mysterious, and is not an "entanglement." Entanglement is something that resembles the story of the two gloves but is more subtle. Here it is.

There are things in nature that we cannot predict. We say they happen "by chance." For instance, if we put a small quantum particle at a point, then the particle starts spontaneously to drift away, but there is no way to predict whether it will drift toward the right or the left. At most, we can anticipate the *probability* that it will move left or right. For instance, we can anticipate that *on average* half of the times the particle will drift to the left and half of the times to the right. But the direction of the first step, we cannot predict.

Entanglement is the fact that the *probabilities* for these unpredictable phenomena can be mysteriously connected for distant events, like the gloves. We can prepare two particles and send one to New York and one to Shanghai, in such a way that each can drift either to the left or to the right, by chance, say half of the times to the right and half of the times to the left, but if the particle in New York drifts to the left, then the particle in Shanghai for sure drifts to the right.

"What's the surprise?" – the smart reader will say – "one particle was naturally 'left drifting' to start with, and the other naturally 'right drifting' to start with, just like the gloves!" But it is not so, and the key point is that this can actually be proven mathematically on the basis of concrete measurements on pairs of particles, thanks to a celebrated theorem proven by the Northern Irish physicist John Bell. The theorem states that if the behavior of the particles was predetermined in this manner, then the probabilities would for sure satisfy certain relations (called "Bell inequalities"). Now the actual measurements have been concretely made in many laboratories around the world and . . . the actual measured probabilities do *not* satisfy the relations written by Bell! In simple words: the particles do not carry with them the information whether they will drift left or right. They sort of "decide" at the last minute what to do, but they decide together, as if they were first consulting one another at a distance!

Today, these measurements are performed by many laboratories, with ever increasing precision and in an ever more clean way. Scientists have been able to produce "entangled pairs" and send them even at great distances. Hundreds of kilometers. Recently, an entangled particle was sent all the way to a spacecraft in orbit and back, and the particle has remained entangled with its sister on earth.

All of these experiments routinely confirm the predictions of quantum theory, which is currently our most reliable theory of the world, and which has been predicting the existence of this mysterious entanglement long before we could actually measure it. The realization that quantum theory predicts this strange phenomenon is due to Albert Einstein and two collaborators: Boris Podolsky and Nathan Rosen. They described it in a celebrated article published in 1935. They did not use the word "entanglement," which first appeared in a letter that the Austrian physicist Erwin Schrödinger wrote to Einstein shortly thereafter (in German, the word he used was *Verschränkung*). In his words: "I would not call [entanglement] one but rather *the* characteristic trait of quantum mechanics, the one that enforces its entire departure from classical lines of thought." The issue remained a bit confused because of the subtleties of joint probabilities, until John Bell nailed it down with its "inequalities": if the particles knew in advance what to decide (as the glove knows if they are left or right), then the probabilities must satisfy his relations. They do not. The particles decide in a consistent manner, with no decision taken in advance.

So, how the hell do the two particles make their decisions consistent? Simple answer: we have no clue.

Could the two particles send a sort of instantaneous message to one another? Hard to believe. If they did, we could perhaps use entanglement to send around instantaneous messages using them. We could use entangled particles like people in the Middle Ages used pigeons to send each other messages: keep couples of entangled particles shared among us and use them to communicate instantaneously when needed. But we can't, because the only way to test entanglement is to compare what happens in both locations: in each single location we have simply a particle that has, say, half probability to drift to the left and half to go to the right. No way to read any useful message into this.

So, there are strange connections between what happens at two different locations, and they cannot be used to send a message from one location to another. They are something subtle that relates the probabilities of what happens on one side with the probabilities of what happens on the other side. They are mysterious connections between things that can be very far apart but still have a strange sort of link.

There are possible technological applications. Not to send better signals, because entanglement is precisely *not* a way to send signals; but in cryptography, namely, in systems that exchange information that should remain private. The reason that entanglement can be used in this way is that the relation between two entangled particles is very delicate: if anything else observes one of the particles, then entanglement is destroyed, and this permits us to verify if a message has traveled unobserved. Entanglement is thus a very concrete reality.

Be careful: entanglement has given rise to some serious hypotheses, but also to all sorts of nonsense about a holistic "connection between anything." One can find all kinds of claims on the Internet: that entanglement explains telepathy, emotional connections, the mind, or anything else. Almost all of this makes no sense. Even worse, there are "quantum medicines" that claim to use entanglement and its mysterious connections to heal the fools who believe anything they are told and pay for it.

Entanglement, indeed, is an extremely delicate phenomenon. It is immediately washed away in our usual macroscopic world, unless the entangled couples are carefully kept isolated from anything else. The mystery of entanglement is not that we do not know *what* precisely happens. We know what precisely happens to two entangled objects: we can predict all relevant probabilities with confidence, and we can find out if two things are entangled or not. What we do not understand is *how* the phenomenon itself is possible, how probabilities can turn out to be related at a great distance, as this seems to contradict our common intuition about the world.

We used to think that everything that happens can only be influenced by what happens right around it. The phenomenon of entanglement questions our basic understanding of what is real. It seems to be telling us that we cannot think that the world is made by objects with their own properties located in space and time. How can something instantaneously influence something else which is far away? Maybe the world is not really as structured in space as it looks? In my own work, quantum gravity, fixed space, and fixed time disappear and are replace by quantum dynamical entities. Could this work explain entanglement? I do not know. Are there subtle connections between different locations in space and time that we had so far missed? We do not know.

Maybe it is something even more fundamental.

PHYSICS AND IMAGINATION

Philip Ball

It would seem perverse, almost rude, not to begin a discussion of imagination in physics with Einstein's famous quote on the topic, voiced during a newspaper interview with the writer George Viereck in 1929:

> I'm enough of an artist to draw freely on my imagination, which
> I think is more important than knowledge. Knowledge is limited.
> Imagination encircles the world.[1]

For a fridge-magnet inspirational quote to celebrate the value of imagination, you need look no further. But context, as so often with Einstein, is everything. He said this after talking about the 1919 expedition led by the British physicist Arthur Eddington to observe the sky during a total solar eclipse off the coast of Africa. Those observations verified the prediction of Einstein's theory of general relativity that starlight would be bent by the gravitational field of a massive body like the sun. Einstein told Viereck that "I would have been surprised if I had been wrong." Viereck, a fascinating figure in his own right, who had previously interviewed (and showed some sympathy for) Adolf Hitler and had written a psychological and gay-inflected Wildean vampire novel in 1907, responded to that supremely confident statement by asking, "Then you trust more to your imagination than to your knowledge?"

You could say that Einstein's reply was a qualified affirmative. And this seems very peculiar, doesn't it, for a "man of science"?

The story dovetails with Einstein's other well-known response to the eclipse experiment. Asked by an assistant (some say a journalist) how he should have felt if the observations had failed to confirm his theory, he is said to have responded, "Then I would feel sorry for the dear Lord. The theory is correct."

Compare that with the statement of another celebrated aphoristic physicist, the American Richard Feynman:

> It doesn't matter how beautiful your [theory] is, it doesn't matter how smart you are. If it disagrees with experiment, it's wrong. That's all there is to it.[2]

Who is right? Einstein trusting to imagination, intuition, and artistry, or Feynman to the brutal judgment of empiricism? If we're talking about scientific methodology, Feynman is right in spirit but nonetheless displaying the limitations of the physicist's common "naïve realist" position about science, which assumes that nature delivers uncomplicated, transparent answers when we put to it questions about our physical theories. Yet Einstein's general relativity was a theory so profoundly motivated and so conceptually satisfying, despite the mind-boggling shift it demanded in conceptions of space and time, that it could not be lightly tossed on the scrapheap of beautiful ideas destroyed by ugly facts.

So, the sensible way to have handled a discrepancy with observed "facts" like those collected by Eddington in his observations of the positions of stars during an eclipse would have been to wonder if the observations were reliable. Indeed, Eddington was later accused of cherry-picking those facts to confirm the theory, perhaps motivated by his Quaker desire to bring about international reconciliation after World War I had triggered the ostracizing of Germany. (It seems that those charges were unfounded.) Nature doesn't lie, but experimentalists can blunder.

There's a deeper reason to valorize Einstein's claim about imagination in physics. What I feel he is really saying is that imagination precedes knowledge, and indeed establishes the precondition for it. You might say that when the shape of imagination sufficiently fits the world, then knowledge results.

We have never needed more reminding of this. In his unfinished magnum opus *Novum Organum,* the seventeenth-century English philosopher Francis Bacon presented knowledge as the product obtained when raw facts – observations about the world – are fed into a kind of science machine (what we might now call an algorithm) and ground into their essence. It was an almost mechanical process: you first collect all the facts you can, and then refine and distill them into general laws and principles about the way the world works. Bacon never completed his account of how this knowledge-extraction process was meant to work, but at any rate no one in science has ever successfully used such a thing, or even knows what it could comprise.

Yet Bacon's vision threatens to return in an age of Big Data – especially in the life sciences, where the availability of information about, say, genome sequences or correlations between genes and traits has outstripped our ability to create theoretical frameworks to make sense of it. There's a feeling afoot not only that data is intrinsically good, but that knowledge has no option but to fall out of it, once the mass of information about the world is large enough.

Physicists have received advance warning of the limitations of that belief. They have their own knowledge machines: sophisticated telescopes and particle detectors, say, and most prominently the Large Hadron Collider and other particle colliders capable of generating eye-watering quantities of data about the interactions between the fundamental constituents of the world. But they already know how little all this data will help without new ideas: without imagination.

For example? There are some good reasons to believe that if physics is going to penetrate still further into the deep laws of nature, it needs a theoretical idea called supersymmetry. So far, all we know about the particles and forces of nature is described by a framework called the Standard Model, which contains all the ingredients seemingly needed to explain everything seen in experiments in particle physics to date. But we know that there's more to the universe than this, for many reasons. For one thing, the current theory of gravity – Einstein's general relativity – is incompatible with the theory of quantum mechanics used to describe atoms and their fundamental particles. Supersymmetry – a putative connection between two currently distinct classes of particles – looks like a promising next step to a deeper physics. Yet so far, the LHC's high-energy collisions have offered no sign that it's true.

What's more, there's nothing in the Standard Model that seems to account for the "dark matter" that astrophysicists need to invoke to explain what they see in the cosmos. This mysterious substance is believed to pervade the universe, invisibly, being felt by ordinary matter only through its gravitational influence. Without something like dark matter, it is hard to make sense of the observed forms and motions of galaxies: how they rotate without shedding stars like a water sprinkler. The reasons to believe in dark matter – and moreover to believe that it exceeds the mass of ordinary visible matter by a factor of about five – are very strong. Yet countless efforts to spot what it consists of have failed to offer any clues. Huge quantities of data constrain the choices, but no evidence supports any of the theories proposed to explain dark matter.

These are – there is no avoiding the issue – failures of imagination. Supersymmetry and dark matter are wholly imagined theories or entities, but the collective imagination of physicists has not yet made them vivid enough to be revealed or disproved. It is possible that this is because they are imaginary in the more literary sense: they exist only in our minds. And they are not alone; dark energy (which causes the universe to expand at an increasing rate) and string theory (one candidate for a theory that would unite gravity and quantum mechanics) are other components of the physicist's imaginarium waiting to be verified and explained or to be dismissed as unicorns, as the ether, as the philosopher's stone.

A single observation – one experiment revealing a discrepancy with a definite theoretical prediction, or one sighting of a new kind of particle – could change the situation. Maybe it will. But it is equally possible that we will need ultimately to concede defeat, and to extend the imagination of physics into new territory: for example, to accept, as some are already arguing, that what we call "dark matter" is a symptom of another physical principle (a modification to the theory of gravity, say) and not a true substance.

There's nothing embarrassing or damning in all this. It's not that physics itself is failing. The situation is just business as usual in science: to have mysteries awaiting explanation, even ones of this magnitude, is a sign of health, not sickness. For individual physicists whose reputations hang (or seem to) on the validity of a particular idea, that's scant comfort. But for the rest of us, it's nothing short of exhilarating to see such deep and broad questions remaining open.

The real point is that imagination in physics is what the paths to the future, to new knowledge, are built from. Actual knowledge – things we can accept as "true," in the sense that they offer tried and tested ways of predicting how the world behaves – has been assembled into an edifice as wonderful and as robust as the Gothic cathedrals of stone, the medieval representations of the physical and spiritual universe. But at the point where knowledge runs out, only imagination can take us further. I think this is what Einstein was driving at.

The invitation is often to suppose that this imagination operates only at the borders of physical theory: at, you might say, the cliff-face of physics that tends to dominate its public image, where we find exotica like string theory, black holes, cosmology, and the Higgs boson. But physics, perhaps more than any other science, has a subtle, fractal-like texture in which gaps in knowledge appear everywhere, at all scales. Imagination was needed to start to understand that strange state of matter made of grains: powders and sand, part fluid and part solid. It is currently blossoming in a field known as topological materials, in which the electrical and magnetic properties are controlled by the abstract mathematical shapes that describe the way electrons are distributed, with twists akin to those in the famous one-sided Möbius strip. It was imagination that has prompted physicists and engineers over the past decade or so to make structures called metamaterials, capable of acting as "invisibility shields" that manipulate and guide light in hitherto inconceivable ways. In all these cases, as in science more broadly, the role of the imagination is not so much to guide us toward answers as to formulate interesting and fruitful new questions.

What does this imagination consist of? We'd do well to give that question more attention. I would suggest that it is, among other things, a way of seeing possibilities: a rehearsal of potential worlds. That's what justifies Einstein's comparison to the work of the artist: as Shakespeare put it, "imagination bodies forth the forms of things unknown." The scientist's theories, as much as the poet's pen, "turns them to shapes and gives to airy nothing a local habitation and a name." That name could be "general relativity" – why not?

What's the source, though? Many ideas in fundamental physics grow from what might seem like the rather arid soil of mathematics. Supersymmetry and string theory are predicated in particular on the conviction that the deepest principles of the physical world are governed by *symmetry*. What this word means at the level of fundamental theory might seem less apparent to the outsider than what it implies in, say, the shape of a Grecian urn or the pattern of wallpaper, but at root it is not so very different: symmetry is about an equivalence of parts and their ability to be transformed one into another, as a left hand becomes a right through the mirror reflection of the looking glass.

Well, it might seem arid, this mathematics. But imagination is as vital here as it is in art. What mathematicians value most in their colleagues is not an ability to churn out airtight proofs of abstract theorems but a kind of creativity that perceives links between disparate ideas, an almost metaphorical way of making connections in which intuition is the architect and proof can come later. Both mathematicians and theoretical physicists commonly speak of having a sense that they are right about an idea long before they can prove it; that proof is "just the engineering" needed to persuade others that the idea will hold up.

Let's be cautious, though, about making "engineering" the prosaic, plodding part of science. The common perception is that theorists do the dreaming and experimentalists just build the apparatus for putting dreams to the test. That's just wrong. For one thing, it's typically experiment that drives theory, not the other way around: it's only when we have new instruments for examining the world that we discover gaps in our understanding, demanding explanation. What's more, experiment too is fueled by imagination. No one tries to see something unprecedented – farther out into space (which means, because light's speed is finite, farther back in time), or into the world of single atoms, or into the spectrum of radiation outside the band of light our eyes can register – unless they have conjured up images of what *might* be there. Sure, you need some existing theory to guide your experimental goals, to show potentially fruitful directions for your gaze; but no one sails into uncharted territory if they think that all they'll find is more of the same, or nothing at all. "If you can't imagine something marvelous, you are not going to find it," says physics Nobel laureate Duncan Haldane. "The barrier to discovering what can be done is actually imagination."[3] And the power and artistry of the experimenter's imagination comes not just from dreaming of what there is to be found in *terra incognita,* but also from devising a means to travel there.

When I speak of dreams, I don't just mean it metaphorically, nor just in the sense of waking reverie. To judge from the testimony of scientists themselves, dreams can function as sources of inspiration.[4] True, we should be a little wary of that; the notion of receiving insight in a dream became a romantic trope in the nineteenth century, and careful historical analysis often reveals some hard and very deliberate graft, as well as a very gradual process of understanding, behind scientific advances that were recast retrospectively as dream revelations. But it happens. Several contemporary physicists have attested to insights that came to them in dreams, as the conscious mind that has been long pondering a problem loosens its bonds on the margins of sleep and admits a little more of the illogic on which imagination thrives.[5]

All the same, we shouldn't think that the physicist's imagination always works in the abstract, in the realm of pure thought. Very often, it takes visual form: finding the right symbolic representation of a problem, such as Feynman's famous "diagrams" for studying questions in the field of quantum electrodynamics (in essence, the theory of how light and matter interact), can unlock the mind in ways that more abstract algebraic mathematics or calculus cannot. Pen and paper can be the fuel of the imagination. As Cambridge physicist Michael Cates (incumbent of the chair previously held by Stephen Hawking and Isaac Newton) has said, "I need a piece of paper in front of me and I'm pushing symbols around on the page . . . so there's this interaction between processing in your head and moving symbols around."[6] Never underestimate the traditional blackboard as a tactile, erasable aid to the imagination. The productivity of such aids is no surprise. Ask a child to think of a story, and it's no easy matter. Give them a doll's house full of figurines, and they're off.

Yet whether it is theoretical or experimental, this imagination in science (as in art) is not idle fantasy. It is a condensation of experience: it takes what you know and plays with it. I do mean "plays": imagination is nothing if not ludic. But it is also the very stuff of thought. One interpretation of cognition, in the context of artificial intelligence, is that it is largely about figuring out the possible consequences of actions we might make in the world: an "inner rehearsal" of imaginary future scenarios. Imagination in science extends that process beyond the self to the world: given that we know *this*, might not things also be arranged like *that*?

It's much more than a guess, then, and as Shakespeare hints, has almost the power of an invocation. Truly, the scientific imagination can invoke into being something that was not there before. Isaac Newton was cautious about his "force of gravity," knowing that he risked (and indeed incurred from his archrival Gottfried Leibniz) accusations of occultism. Yet all the same, this "force" became – and remains – a "thing" in physics, even if we can regard it as a figure of speech, a convenient conceptual tool that Einstein's general relativity now invites us to regard otherwise as curvature of space-time. It's a process entirely analogous to the way Shakespeare goes on to speak, in *A Midsummer Night's Dream,* of how, viewing some event or effect, we can't resist imagining a cause:

> Such tricks hath strong imagination,
> That if it would but apprehend some joy,
> It comprehends some bringer of that joy.

In this way we're reminded that imagination shares an etymology with "magic" – which, in the age just before the time of Isaac Newton, did not necessarily mean superstitious agency but the "hidden forces" by which natural magicians comprehended and claimed to manipulate nature. According to one of the most eminent of these pre-Newtonian natural magicians, the Swiss physician Paracelsus, imagination as a mental activity could have a real causative influence on the physical world: it brought things and effects into being.

In this regard, then, Newton wasn't, as John Maynard Keynes claimed, the "last of the magicians" because of his belief in occult forces (such as gravity, acting invisibly across space). No, if *that* was Newton's "magic" then today's physicists share a conviction in it, for any model in physics awards imagination this role of employing imagined causative agencies – things we might not perceive directly but which manifest through their effects, such as dark matter, dark energy, or the Higgs field – to explain what we see.

Now, though, physics places demands on the imagination as never before. I'm struck by how dark matter and dark energy, say, commandeer familiar concepts (mass, energy) that may or may not turn out to be appropriate to the occasion. Even more challenging are efforts to provide some physical picture of quantum mechanics, the kind of physics generally used to describe atoms and fundamental particles. These objects don't seem to conform to our intuitions derived from the everyday world of rocks and stones, tennis balls and space rockets. They can, for example, sometimes display behavior that we associate not with particles but with waves. They appear to be able to influence one another instantaneously over long distances; they are said to exist "in several states or places at once."

Yet these descriptions are attempts – often clumsy, sometimes misleading – to make quantum mechanics fit into the forms of our conventional "classical" imagination. Arguments and misperceptions follow, or a disheartening decision to draw a veil over quantum improprieties by calling them "weird." We can and should do better, but this will require a reshaping, an expansion, of our imaginative faculties. We have to develop a kind of intuition that is not constrained by our daily experience – because if there's one thing we can be sure about in quantum mechanics, it's that it demands the possibility of phenomena that lie outside this experience.

To venture into unknown territory, where imagination is at a premium, is a risk. To put it bluntly, your imagination is more likely to lead you astray than toward the truth. It is no magical guarantor of insight. Will you take that risk? Mathematical physicist Jon Keating has put the problem succinctly: "[How can we] encourage people to make them feel more comfortable with the failure that comes with most creative and imaginative ideas?"[7] Unless we get better at that, educationally or institutionally, science will suffer.

And it's very possible that physicists won't alone accomplish the feats of imagination needed to crack their hardest problems. They may need to find inspiration from philosophy, art, literature, or aesthetics. Imagination doesn't recognize categories and boundaries – it is a power that aims to encircle the world.

[1] George Sylvester Viereck, "What Life Means to Einstein," *The Saturday Evening Post*, October 26, 1929 (Indianapolis, Indiana), p. 117.

[2] Richard Feynman, lecture at Cornell University, 1964, https://www.youtube.com/watch?v=OL6-x0modwY (all URLs accessed in January 2019).

[3] Duncan Haldane, in "Princeton professor says Nobel physics win 'gratifying,'" *Reuters*, https://reuters.screenocean.com/record/84768.

[4] Philip Ball, "Chemistry in soft focus," *Chemistry in Britain* 38, no. 9 (2002), p. 32.

[5] Philip Ball, "In search of claritons," *Physics World* 29, no. 10 (2016), p. 30.

[6] Rosie Robison, report on "Physics Imagination Retreat: Exploring the Role of Imagination in the Work of Scientists," June 29 to July 1, 2016, p. 9, Imagination Institute, University of Pennsylvania, https://cpb-us-w2.wpmucdn.com/web.sas.upenn.edu/dist/8/471/files/2018/06/Physics_Imagination_Retreat_Report-wl9qnd.pdf.

[7] Ibid., p. 20.

UNRAVELING SURREALISM'S ENTANGLEMENT WITH PHYSICS

Gavin Parkinson

In 1959, the former Surrealist painter Salvador Dalí proposed a reading of the painting by his friend Marcel Duchamp, *The King and Queen Surrounded by Swift Nudes* from 1912, as a snapshot of the underlying structure of matter:

> In painting *The King and the Queen Traversed by Swift Nudes* [*sic*], the genius of Marcel Duchamp proclaimed nothing less than the notarial act of the new intra-atomic structure of the universe, that is, the discontinuity of matter. In fact, the king and the queen can be traversed by swift nudes because matter is discontinuous. It is easy to understand that swift nudes are indivisible bodies, the corpuscles, the charged elementary particles of quantum physics . . . .[1]

Dalí's was a typically idiosyncratic account of the metaphorical language of this painting. It is true that Duchamp had been a reader of the great French mathematician Henri Poincaré and interested in the various popular and scientific notions of the fourth dimension before World War I, but he had no knowledge of the developing theory of quanta. It is more likely that the stated iconography of *The King and Queen Surrounded by Swift Nudes* was drawn from Duchamp's love of chess, and its monochrome and semi-abstract style was certainly sourced in Cubism. However, if Dalí's reading of the painting had less to do with Duchamp's preoccupations in 1912 than his own in 1959, then those dates serve a useful purpose as a period frame for an examination of Surrealism's engagement with physics.

Duchamp had been an important figure of the preceding generation for young members of Paris Dada like the poet André Breton, who would soon become the main theorist and spokesperson of the Surrealist group. His *Manifesto of Surrealism* was published in 1924, and Breton had been aware as early as 1922 of Duchamp's work-in-progress on his masterpiece *The Bride Stripped Bare by Her Bachelors, Even* or *The Large Glass* (1915–23).[2] The *Glass* has now been thoroughly historicized in terms of Duchamp's knowledge of the science and technology of the era, which Breton was also keenly attuned to, partly through the counsel of the older poet Paul Valéry, his mentor from 1914 until about 1921 and a friend of Marie Curie, Jean Perrin, Paul Langevin, Niels Bohr, and Louis de Broglie.[3] Breton owned the 1921 French translation of Albert Einstein's semi-popular exposition *Relativity: The Special and the General Theory,* which alongside his access to Valéry would have served as a solid introduction to relativity.[4]

The language of relativity is poeticized in some of Breton's writings of the time and it influenced his perception of art. This is confirmed in a letter that Breton sent to the painter André Derain in 1921, where he wrote of his interpretation of Max Ernst and his recent work: "[s]ometimes I claim that we owe a pictorial language to him in accordance with the new conception of things advanced by Einstein."[5] Here André Breton was alluding to the short text he wrote to accompany Ernst's first exhibition in Paris in May 1921, where Breton's handling of relativity was determined by the absurdist spirit of Dada.[6] Voicing his admiration for the imagery of Ernst's collages and paintings, he compared their effect on the viewer of *dépaysement* (disorientation) – a category that would become fundamental to Surrealist poetics – to Einstein's description of the "separate realities" experienced by observers moving relative to each other:

> Belief in an absolute time and space seems about to disappear. Dada does not claim to be modern. It judges futile, as well, the subjection to laws of a given perspective. But the marvelous power to bring together, without leaving the field of our experience, two distant realities and from their reconciliation to draw a spark; to put within the grasp of our senses abstract figures endowed with the same intensity, the same relief as the others; and, by depriving us of our frames of reference, to disorient us within our memories, that is what preoccupies Dada for now.[7]

Here, the production of the emotive spark – the central mechanism of what would become Surrealism's theory of the poetic merveilleux – is identified at the time of its first brief flaring with the new physics of space-time.

Comparable to the theory of entanglement, Surrealism's poetic theory insists on the mysterious relationship that exists between distant entities. It was articulated as follows by Breton in the *Manifesto:* "It is, as it were, from the fortuitous juxtaposition of the two terms that a particular light has sprung, *the light of the image,* to which we are infinitely sensitive. The value of the image depends upon the beauty of the spark obtained; it is, consequently, a function of the difference of potential between the two conductors."[8] In this formulation, Breton sought to replace "the realistic attitude, inspired by positivism" with the poet Pierre Reverdy's 1918 formulation of the poetic image, also quoted in the *Manifesto:* "The image is a pure creation of the mind. / It cannot be born from a comparison but from a juxtaposition of two more or less distant realities. / The more the relationship between the two juxtaposed realities is distant and true, the stronger the image will be – the greater its emotional power and poetic reality . . . ."[9]

Consistent with this, Breton would much later extol "the spontaneous, clairvoyant, insolent connection established under certain conditions between two things whose conjunction would not be permitted by common sense" in the important postwar reevaluation of poetics, and analogy specifically, titled *Ascendant Sign* (1947).[10]

Dalí's long-standing knowledge of the imagery of artists like Ernst made his initiation into the Surrealist group in 1929 fairly straightforward. During the period that he entered the Surrealist circle, his work drew quite consciously upon some of the stranger concepts of relativity that Breton had been exposed to several years earlier as he sketched in the background of Surrealist theory. Popular expositions of relativity – like the English physicist Arthur Eddington's – reveled in their description of a strange new world effecting changes to physical bodies that are invisible to the naked eye. However, the temptation to exaggerate them by raising them to the visible, human scale is compelling. Such was the case in popular imagery, too, where both relativity and Surrealism were thought of as creating an "alternative terrain" or "world" of their own. Dalí's attraction to Surrealism, along with his fascination, evinced in his writings, with the phenomena described by relativity – such as the constant variations taking place in the forms of physical bodies, on however small a scale – helped persuade him to import and amplify the elastic, contorted world of relativity into his early Surrealist canvases as soft, viscous, attenuated forms, sometimes suggestive of anamorphosis, as in his *Female Nude* and *Bather,* both of 1928.

His paintings of the late 1920s demonstrate that Dalí sought a language in modern physics that would enable him to depict this "real" world. By 1934, in the pages of the Surrealist journal *Minotaure,* he was insisting upon relativity's fourth, temporal dimension as a veritable window onto the surreal, describing it as the "delirious and Surrealist dimension par excellence."[11]

By then, there was a greater understanding, through popular writings and the philosophy of science, of the equal revolution in the understanding of the physical world engendered by quantum mechanics, which was formulated more or less alongside the birth of Surrealism in the mid-1920s. This is demonstrated, also in 1934, by the explicit case made by Ernst for a philosophical link between Surrealism and quantum theory. Ernst had become acquainted with the "amplifier theory" of the young physicist Pascual Jordan, which argued that on the Newtonian, observable level, "the statistical acausality of atomic reactions" averages out to a basic causality for inanimate matter, whereas for animate matter, "the acausality of specific atomic reactions amplify to a macroscopically operative acausality."[12] For Jordan, the free will that characterizes humans, animals, and plants is a direct echo of the same acausal constituents that form inanimate matter and yet disappear for the latter when raised to the macroscopic level. Jordan believed that his "amplifier theory" gave quantum mechanics a general application, extending to biology and psychology. Gentle disagreements with Bohr did not prevent him from publishing on the subject in 1932 in Arnold Berliner's interdisciplinary journal *Die Naturwissenschaften.*

This is where Ernst came into contact with Jordan's writings, which he would mobilize in his brief essay titled "Was ist Surrealismus?" (What Is Surrealism?), published in 1934. Ernst attempted to answer that question by rethinking the role of the artist-observer in the creative process, separating Surrealism from traditional models of philosophical inquiry and attuning its practices and findings to those of quantum mechanics. Ernst's experimental technique of frottage creates suggestive, patterned surfaces from rubbings made with pencil or charcoal over wood, which are then identified and actualized through additional elements applied by Ernst, in the same way a Rorschach ink blot can be visually interpreted initially and then worked up so that it conforms to that interpretation. Alongside other experimental methods devised by Surrealism, he commends the technique as evidence of "an exploration of a new and incomparably wider field of experience in which the borders between the so-called interior and exterior worlds (as defined by classical philosophy) are becoming increasingly blurred and will probably disappear altogether one day."[13]

Pressing frottage's claims to innovatory relevance, Ernst continued as follows:

> The revolutionary significance of this initially absurd-seeming description of nature is pointed up by the fact that analogous results have been achieved in modern microphysics. After measuring the path of an electron moving without application of external force, and then measuring its location, P. Jordan concluded: But this distinction [between exterior and interior world] is robbed of one of its main props by the experimental refutation of the idea that facts are present in the exterior world that have an objective existence independent of the process of observation.[14]

In other words, like the physicists exploring quanta, the Surrealists were charting a zone that could no more be satisfactorily designated "objective" than "subjective." The acausality of both unconscious and subatomic phenomena had forced a breach with the classical descriptive rationale served by those unitary and mutually exclusive categories. Ernst goes on to say of Surrealist painters:

> they move freely, daringly and naturally on the physically and mentally quite real ("surreal"), if still largely undefined, frontier between interior and exterior world, registering what they see and experience there, and intervening where their revolutionary instincts suggest. In fact, the fundamental opposition between meditation and action (again, according to the view of classical philosophy) disappears along with the fundamental distinction between exterior and interior world, and the universal significance of Surrealism lies precisely in the fact that, in the wake of this discovery, no area of life can remain closed to it.[15]

Ernst spoke from a long-standing knowledge of psychoanalysis and obvious interest in the "irrational" aspects of nature, which he often contrasted with the mechanized rationalism of modernity.

The Chilean artist Roberto Matta and the Austrian painter Wolfgang Paalen joined the Surrealist group in the mid-1930s as Dalí was falling into disfavor. Both were already enthusiastic about the potential that was offered to the imagery of modern painting by the "world" described by subatomic physics. Matta's oeuvre is marked by his contact with Duchamp, whom he had known in Paris and with whom he became reacquainted during World War II when Duchamp arrived in New York in 1942 (Matta mentioned later the "weekly lunches" they shared until 1944).[16] The poet Octavio Paz later wrote of the period: "in Surrealist painting a radical metamorphosis was in progress, initiated by Duchamp and realized by Matta, a fusion of eroticism, humor, and the new physics."[17] While the "Freudian" title of *The Vertigo of Eros* suggests that Matta wanted the painting to be thought of as an infinite, chaotic, unmapped "mindscape," the hovering particle-like objects orbiting circular scratches on the surface of the canvas are also reminiscent of the subatomic "world," which physics had found to be composed mainly of empty space, described most engagingly by Arthur Eddington. In both paintings and interviews, Roberto Matta was deliberately imprecise in his deployment of the resources of modern physics in their intersection with other fields of knowledge. In fact, the invention of a new imagery of the invisible was his main goal.

Of all the Surrealists, Paalen was perhaps the most conversant on quantum mechanics, from which he developed a pictorial language during the war while living in Mexico. The ripples that pulse across the surfaces of his canvases of those years bear witness to his pursuit of a nature understood not in the forms which it ordinarily inhabits in landscape painting, but rather in its behavioral, vibratory sense – the subatomic flux of nature as revealed by the wave mechanics of Louis de Broglie and Erwin Schrödinger. Typical is *Figure pandynamique* of 1940 to which Paalen attached some importance. A multitude of whirling forms seem to emerge from rings, possibly meant to suggest holes leading to dimensions beyond conventional space. The manner in which these rings increase or decrease in size along a single axis to create mini cyclones suggests a throbbing motion, while the repeating rings sometimes end at their "tail" with a sphere representing, presumably, a particle. Works like these issued from Paalen's bid to produce an artistic theory to compete with Breton's Surrealism from which he had become increasingly alienated in the 1940s. They were aligned less in Paalen's mind with the Cubism that Guillaume Apollinaire, the poet and friend of Pablo Picasso, had associated with relativity than with progressive trends like the abstraction of Wassily Kandinsky, which

Paalen linked to the passing of the dominance of Newtonian mechanics and the ascendance of quantum theory, heralding, he believed, a new age.

The positive reception afforded physics by Surrealism continued for at least two years after the war and the advent of the atomic age brought about by the explosions over Hiroshima and Nagasaki. From 1948, the Surrealists formulated a critique of nuclear physics and the bomb that converged with their anti-authoritarian rhetoric. This culminated in the 1958 broadside "Expose the Physicists, Empty the Laboratories," defining Surrealism's position against atomic energy in the period of the Cold War.[18] Dalí's escalating fascination with physics in the era of nuclear energy no doubt furthered this distrust, since he had left Surrealism behind at the beginning of the war and was now thoroughly reviled by the Surrealists as "Avida Dollars," the avaricious Catholic and admirer of General Franco. In fact, Dalí's most recent exhibition at the M. Knoedler & Co. Gallery in New York, including the series of diffusive drawings and watercolors collectively titled *Anti-Matter Archangels and Celestial Visions,* had opened on February 18, 1958, the same date appended to the Surrealists' attack on physics.

Dalí's mélange of science and religion received another outing later in 1958 when his "Anti-Matter Manifesto" appeared as the preface to the catalogue of another exhibition of his work, at the Carstairs Gallery in New York.[19] Speaking of his love for physics, Dalí made his now well-known post-Surrealist conversion statement, redrawing his intellectual genealogy along lines laid down by quantum mechanics rather than psychoanalysis:

> In the Surrealist period I wanted to create the iconography of the interior world – the world of the marvelous, of my father Freud. I succeeded in doing it.
>
> Today the exterior world – that of physics – has transcended the one of psychology. My father today is Dr. Heisenberg.
>
> It is with pi-mesons and the most gelatinous and indeterminate neutrinos that I want to paint the beauty of the angels and of reality.[20]

In the postwar period, the Surrealists did not really need another reason to hate Dalí, but he was more than happy to give them one. It is difficult to conceive of a starker contrast than the one lying between Surrealism's support for the Comité de Lutte Anti-Nucléaire with their call to "boycott conferences devoted to the exaltation of the atom" and Dalí's exactly contemporary, provocative celebration and mythification of its powers.[21]

Dalí's particle angels found no takers among the Surrealists, of course, but his current work with the odd analogy it drew, to say the least, between Catholicism and antimatter did receive a favorable response from one quarter. The Surrealists' old friend Marcel Duchamp had been tasked with coordinating their exhibition *Surrealist Intrusion in the Enchanters' Domain* at the d'Arcy Galleries in New York, codirected with Breton back in Paris and set to open in November 1960. Courtesy of unusual indelicacy on the part of Duchamp, who over the years had pulled off the unique trick of juggling friendships with both Dalí and Breton at the same time, one of the star exhibits from the Carstairs Gallery – the large-scale canvas completed that year titled *The Sistine Madonna* – made its way into *Surrealist Intrusion.* The Surrealists were infuriated not just by the inclusion in their own exhibition of a painting by "Hitler's former apologist, the fascist painter, the religious bigot, and the avowed racist," as they put it in their protest "We Don't Ear It that Way," but by the presence in their show of a work with such a theme when their confidence in physics as a force for good had all but expired.[22] The alliance with Duchamp would just about recover over the next few years, but the one with physics would be as lifeless from this point as their former comradeship with Dalí.

[1] Salvador Dalí, *The King and the Queen Traversed by Swift Nudes* (1959), in *The Collected Writings of Salvador Dalí*, ed. and trans. Haim Finkelstein (Cambridge: Cambridge University Press, 1998), pp. 367–69, esp. p. 367.

[2] André Breton, "Marcel Duchamp" (1922), in *The Lost Steps*, trans. Mark Polizzotti (Lincoln, NB, and London: University of Nebraska Press, 1996), pp. 85–88, esp. p. 88.

[3] Linda Dalrymple Henderson, *Duchamp in Context: Science and Technology in the Large Glass and Related Works* (Princeton, NJ: Princeton University Press, 1998).

[4] Albert Einstein, *La théorie de la Relativité: Restreinte et Généralisée*, trans. J. Rouvière (Paris: Gauthier-Villars et Cie, 1921).

[5] "Je prétends quelquefois qu'on lui doit une figure plastique conforme aux nouveaux modes des réprésentation mises en vigueur par Einstein." Letter of October 3, 1921, quoted in Werner Spies, *Max Ernst, Collages: The Invention of the Surrealist Universe*, trans. John William Gabriel (New York: Harry N. Williams, 1991), p. 256, note 333.

[6] Breton, "Max Ernst" (1921), in *The Lost Steps*, pp. 60–61.

[7] Ibid. (translation modified).

[8] André Breton, "Manifesto of Surrealism" (1924), in *Manifestoes of Surrealism*, trans. Richard Seaver and Helen R. Lane (Ann Arbor: University of Michigan Press, 1972), pp. 1-47 and 36.

[9] Breton, *Manifestoes of Surrealism*, p. 20 (Breton's italics).

[10] André Breton, "Ascendant Sign" (1947-48), in *Free Rein*, trans. Michel Parmentier and Jacqueline d'Amboise (Lincoln and London: University of Nebraska Press, 1995), pp. 104-7, esp. p. 104.

[11] Salvador Dalí, "Aerodynamic Apparitions of 'Beings-Objects,'" (1934), in Finkelstein, *Collected Writings*, pp. 207-11, esp. p. 209. See Salvador Dalí, "Apparitions aérodynamique des 'Être-Objets,'" *Minotaure* 6 (Winter 1934-35), pp. 33-34.

[12] Unpublished manuscript translated and quoted in Finn Aaserud, *Redirecting Science: Niels Bohr, Philanthropy, and the Rise of Nuclear Physics* (Cambridge: Cambridge University Press, 1990), p. 84.

[13] Max Ernst, "What is Surrealism?" (1934), in *Max Ernst: A Retrospective*, ed. Werner Spies (London: Tate Gallery, 1991), pp. 309-10, esp. p. 310.

[14] Ibid., p. 310. I am very grateful to Simone Perks, who alerted me to Ernst's text quoting Jordan. Ernst quoted the final sentence from the following passage by Jordan: "Diese Erschütterung der klassisch-philosophischen Vorstellung des von der Beobachtung unberührten Objektes bedingt übrigens eine gewisse Abmilderung eines Gegensatzes, der von fast allen früheren Philosophen als ein *absoluter*, grundsätzlicher und vollkommener betrachtet wurde; wir meinen den Gegnesatz zwischen objektiver Beobachtung und rein subjektivem Erleben. Man ist gewohnt, von der Betrachtung und Beobachtung der sog. 'Außenwelt' die Beobachtung der eigenen subjektiven 'Innenwelt' grundsätzlich zu unterscheiden. Aber dieser Unterscheidung wird eine Hauptstütze entzogen mit der experimentellen Widerlegung der Vorstellung, daß in der 'Außenwelt' Tatbestände vorlägen, welche unabhängig vom Beobachtungprozeß ein 'objektives' Dasein besitzen." Pascual Jordan, "Die Quantenmechanik und die Grundprobleme der Biologie und Psychologie," *Die Naturwissenschaften* 20 (November 4, 1932), pp. 815-21, esp. p. 818.

[15] Spies, *Max Ernst*, p. 310.

[16] Matta, "Auto-Elasto-Infra-Biography," in Matta: *Entretiens Morphologiques: Notebook no. 1, 1936-44*, ed. Germana Ferrari (London: Sistan, 1987), pp. 12-19, esp. p. 16.

[17] Octavio Paz, "Vestibule," in *Crosscurrents of Modernism: Four Latin American Painters*, ed. Valerie Fletcher (Washington, DC: Smithsonian Institution Press, 1992), pp. 274-81, esp. p. 277.

[18] The Surrealist Group, "Démasquez les physiciens, videz les laboratoires," in *Tracts surréalistes et déclarations collectives, 1922-1969*, ed. José Pierre, vol. 2: *1940-1969* (Paris: Le Terrain Vague, 1982), pp. 172-73 and 377-78.

[18] Dalí, "Anti-Matter Manifesto," in Finkelstein, *Collected Writings*, pp. 366-67.

[20] Ibid., p. 366.

[21] The Surrealist Group, "Expose the Physicists, Empty the Laboratories," in *Surrealism Against the Current: Tracts and Declarations*, ed. and trans. Michael Richardson and Krzysztof Fijalkowski (London: Pluto Press, 2001), pp. 164-65, esp. p. 165.

[22] The Surrealist Group, "We Don't Ear It that Way," in Pierre, *Tracts surréalistes*, vol. 2, p. 210.

THE LIVE CREATURE AND ETHEREAL THINGS

Nicola Triscott

When the philosopher John Dewey wrote about *Art as Experience* in 1934,[1] he was attempting to shift the understanding of what is important and characteristic about art from its physical manifestations in the object to the process in its entirety, a process whose fundamental element is no longer the material "work of art" but rather the development of "an experience," such an experience being something that personally affects your life. He used the term "ethereal things," which he took from John Keats, to condemn aesthetic theories that elevated art too far above its pragmatic experiential roots. In a similar spirit to Dewey's critique of art theory in the 1930s, I want to broaden the framing of contemporary physics from a methodology and field of knowledge to a process, an exchange between the material and the "live creature" – the human, and a practical, experienced reality.

Astrophysics and particle physics probe the limits of the known, the limits of language and concept, and present a kind of abstraction or transcendence that can free us from commonsensical thinking. Yet they are undertaken by human beings in real-world laboratories and observatories, driven, shaped, and influenced by those people's cultural backgrounds, education, funding requirements, teaching obligations, family commitments, societal norms, aspirations, rivalries, and dreams. How do these human beings co-create knowledge about the structure and mechanics of the universe, and how is this process and knowledge put forth into the world? These questions are intertwined because the way in which physics represents itself to the non-expert tends to focus on facts and knowledge, rather than on the ways and whys and wherefores by which it is produced. This can lead to a gap in the cultural imaginary and social understanding of physics, a gap between an awareness of these people called physicists and this set of knowledge called physics, in which there is little sense of how physics unfolds through its processes, personalities, and places, or of how one thing leads to or affects the other. Within this gap, as Kevin J. Pugh and Mark Girod similarly note of science education that focuses solely on understanding scientific concepts, we can also lose awareness that science has the potential "to enrich everyday life, vitalize experience, and provide us with aesthetic satisfaction."[2]

Given current concerns around science recruitment and public belief in and support for science, it is timely to address this gap. In this essay, I will approach the subject of bringing the physics into the human experience – or bringing the human back to the center of physics – from two angles, one a discussion of the culture of physics, questioning the notion of a "no culture" of physics, the other a consideration of the expression of physics as a human, material activity into wider society. These two perspectives broadly parallel those of the artist and physicist contributors to my book.[3] In addition, I will interweave some thoughts on the value of artistic practice to physics itself and the wider cultural expression and societal role of physics.

The Culture of Physics

To posit the existence of a "culture of physics," and to suggest that culture might shape how physics is done, can trigger polarized positions. There are those who think that the social environment, belief systems, and behaviors of the physics community must influence what science is pursued and how, while others insist that the endeavor of physics is not affected by its culture, or that it is culture-free.[4] The so-called "science wars"[5] are a sad example of binary position-taking, in which scientific realists clashed with postmodernists, whom they regarded as rejecting scientific objectivity, the scientific method, and scientific knowledge. This is a great shame, because recognizing and enabling studies of the culture of physics could potentially shed light on some of the issues that concern scientists, those around the recruitment, retention, and diversity of scientists, and those concerning public support for science. In addition, enabling further porosity with other cultural influences and disciplinary perspectives could potentially bring benefits to science itself, in terms of alternative perspectives and methodologies and new ideas, which seem entirely in tune with science's pursuit of inquiry into how the universe works. The position of some scientists that nothing cultural (and certainly nothing artistic) could possibly affect the scientific method seems unlikely, given that it is humans who create and re-create the scientific method. There is no one immutable method, and the scientific method tends to evade simple definition, although there are principles in common around testing and replicability.

Many sociological and anthropological studies have shown that physics is a culture with internal beliefs and value systems, and that there are differences in cultures of science across different parts of the world, reflecting varied histories, societal attitudes, institutional contexts, and dominant social makeup. Yet, many scientists still prefer to believe that their world is basically independent of culture, "a culture of no culture," as anthropologist Sharon Traweek terms it, "which longs passionately for a world without loose ends, without temperament, gender, nationalism or other sources of disorder."[6] Despite this longing (or ideal), gender and cultural background do seem to matter in the self-selection, institutional selection, and retention of those who enter into the culture of physics.

In Cathrine Hasse's analysis of connections between gender and physics,[7] she notes that, in general, the natural sciences report increasing problems with recruitment, especially of female physics students. Notably, she finds that this is primarily in the northwestern world, where there is also a disproportionate leakage of women from scientific careers at every stage in the academic hierarchy, which challenges the notion of an internally self-regulating, culture-free science. Attempts to explain this away in terms of restrictions in educational or employment opportunities for women in the wider culture or childcare responsibilities are belied by puzzling differential patterns, where the proportion of female physicists is found to be higher the further south and east we go, not just in Europe but across the globe. In her fascinating study, in which she interviewed physicists from different European countries, Hasse found a connection to physics that had nothing to do with gender (indeed the Italian female physics students did not regard physics as a "gendered" subject at all) but rather concerned the relationship between humanities and natural sciences in a local, cultural-historically shaped context.

It was this cultural contrast that gave Hasse new insight into the gender differences found. In Italy, she discovered, it was possible to study physics at a university level with a background in the humanities: ". . . many of the very successful Italian professors in physics that I interviewed turned out to have entered their study with a high school background as classical studies. They were versed in philosophy and cultural history, studying Aristotle in Greek, reading Cicero in Latin and the like,"[8] whereas in Denmark, as in the UK, humanities and natural sciences are generally seen as mutually exclusive. Indeed, in Italy a background in classical studies was seen as an asset to a physicist. Hasse mentions a professor in physics in Italy, who came to physics from a scientific background, explaining:

> I myself have a background equivalent to a mathematical-physicist high school student in Denmark. But here in Italy it is rather an advantage to have a classical linguistic background, when you start physics studies at university level. The "classical" students are simply better at analyzing. What we learned in science high school was to think much more "mechanically" – to think in the correct answers. I have always believed students with a classical background are the most advantaged.[9]

Further, Hasse notes, her interviews with Danish, Swedish, and English female physicists showed a much more problematic relationship between being female and being a physicist than with the Italians. Many reported that through their whole career they had "felt like outsiders."[10] In these northwestern European cultures, she remarks, people have come to understand a connection between physics and "hard science," and to connect hard science with being male. These connections are not made when you move south and east.

This suggests that not only gender matters in the selection of those who pass into physics, but so do personality traits and cultural and creative interests. Suchitra Sebastian, a condensed matter physicist at Cavendish Laboratory, University of Cambridge, makes a similar point:

> People in the field assume that to be a good physicist you have to be "like them." I think in their mind they're not selecting for anything other than good physics, but of course in reality they're selecting for a plethora of characteristics that are to do with the person and not just the physics that they do.[11]

Sebastian earned her undergraduate degree in physics at a liberal arts college and says that her multidisciplinary outlook conflicted with what was expected of her when she became a physicist. Moving into physics as a profession, she felt huge pressure to be "a certain kind of physicist," to have gravitas and to be interested only in physics, and therefore she separated out the parts of her personality that didn't fit. It was only when she began to engage with the public that she realized they simply weren't interested in hearing about the science she was doing when she spoke about it in the abstract. They sensed she wasn't being authentic, and it was then that she began to reconnect her other interests with her science.

In 2018, I edited a book with the artist Fiona Crisp called *The Live Creature and Ethereal Things: Physics in Culture*, which followed a workshop that we convened with the Institute of Physics in London, in which Sebastian participated. She and the other physicists at the workshop who were invited to contribute to the book were asked to bring their individual authentic selves to their writing: to write about science in a way that only they could. This produced a series of insightful and compelling contributions that weave connections between their scientific selves and their cultural selves. In her text, particle physicist Tara Shears expresses a personality trait that perhaps makes her an easier "fit" for the way the physics world is at the moment: a joy in simplicity and in finding a way to order the world. She notes that "particle physics is full of people like me." She writes of her love of mathematics, but she also writes of her love of literature, poetry, and art. She feels drawn, she says, to their economy and power of expression, as much as that of mathematics. Flaviu Cipcigan directly relates his passion for dance to physics, arguing that we need an expansion of how we represent knowledge from something that exists in our heads to environments that use all of our senses. Cosmologist Mark Neyrinck discusses his fascination with origami as a tactile, topographical, and visual way to explore cosmic structure. Marek Kukula reflects on a history of art powering new perspectives about science and considers its relation to disruptive thinking, a characteristic that, he suggests, the paradigm-breaking astronomer Galileo inherited from his father, the musician and composer Vincenzo Galilei.

In a conversation between the dark matter physicist Chamkaur Ghag and the artist Ansuman Biswas, they discuss embodied knowledge, starting from a shared interest in meditation – a "contemplative science." They critique the culture of "nonstop thinking" in physics, noting that "the speed at which things are progressing is an indication that there's very little reflection." Ghag calls for scientists to slow down, come out of their heads, and ask themselves where science is going. He suggests that the next evolution of the scientific method requires putting science back into context, and considers that such an approach would be welcomed by many scientists, although the established culture and structures of physics make this challenging.

I sense that the conversation between Ghag and Biswas represents an emerging new mode of science, in which the skills and interests of people inside and outside science are starting to come together to create new ways for people to engage with science and for scientists to develop a new ethical relationship with the world.

Physics in Culture

Having outlined some of the arguments for recognizing that physics is a human experience, with its own culture, and how this recognition can benefit the physics community, in this section I will explore some of the urgencies for physics to connect to wider culture and society, and how approaches and insights from artistic practice are helping to create these new connections as well as contributing to science itself.

The astronomer Roger Malina remarks, in *The Live Creature and Ethereal Things: Physics in Culture,* that one of the problems for modern science is that it doesn't make sense on the human scale. Almost all physical knowledge about the world and the universe now comes through scientific instruments and is pursued at scales and distances far beyond our human sensing. In fact, he adds, much science is done today by experiments on data rather than experiments on the world. As more and more science comes to us through this mediated data, there are cultural implications in the disconnect between the science and the intimate world that we live in every day. This disconnect is a problem because it is important to be able to know which scientific facts are useful and which are not useful, and this requires that scientific knowledge is assimilated by society and contextualized.[12] He refers to this process as "making science intimate." While scientists do this, for example, creating simulations that mimic how the world works, he considers that artists increasingly play an invaluable role in this process, as do movements such as citizen science and people's science. Bringing artists into the scientific realm, he argues, can lead to better science, and it can also lead to different science through the process of embedding science into wider society.

The contribution that artists can make to science is increasingly recognized by scientific institutions, both in terms of what artists bring to science itself and the scientific environment, and of developing a more fluid relationship between science and society. More and more science institutions are setting up schemes designed to bring artists into the realm of science, not as illustrators but to bring their different perspectives and practices to work alongside and inform those of science. Perhaps the most long-standing of these is the Visiting Artists program at the Massachusetts Institute of Technology (MIT). Another well-known example is Arts at CERN, the official arts program of CERN, the large particle physics research facility in Switzerland.

One of MIT's artists-in-residence was Tomás Saraceno. An artist trained as an architect, Saraceno deploys insights from engineering, physics, chemistry, aeronautics, and materials science in his work. During his MIT residency, the artist investigated the intricate geometry of spiderwebs and developed an original method to scan a three-dimensional web. Learning that spiders are known to use many astronomical cues for navigation, including the position of the moon and patterns formed by polarization of light in the sky, Saraceno was inspired to produce a series of works, in collaboration with arachnologists, astrophysicists, architects, and engineers, involving live spiders spinning universes. In these spiderweb universes, intricate filaments allude to dwarf and spiral galaxies, nebulae, and quasars. His largest work, *How to Entangle the Universe in a Spider Web,* involved 7,000 spiders in its creation.

The artist group Semiconductor (Ruth Jarman and Joe Gerhardt) has undertaken many residencies in scientific institutions around the world, including at CERN. Their film *Do You Think Science,* made during a residency at the NASA Space Sciences Laboratory in California, asks several space physicists an unheard-of question about the limits of scientific understanding, prompting most to pause and then falter their way through unfamiliar metaphysical terrain. One just laughs: "Everything? No!," then asks in bafflement: "Do people actually say yes?" Other works by Semiconductor, such as *Brilliant Noise* and *Black Rain,* mine raw unprocessed scientific data to create mesmerizing static-filled, jittery, black-and-white films of solar storms, creating a radically different aesthetic from the hyper-colored, cleaned up images of cosmic objects, such as those created from Hubble Space Telescope data, with which we are more familiar.

The scientific method is not a static thing. It continues to evolve; indeed, it could be said to need a redesign with the development of sciences of complexity and artificial intelligence. The work of artists, such as Saraceno, Semiconductor, and many others who work in the terrain of science, can help to inform scientists' intuitions, vocabularies, and the way that they conceptualize and describe things, as science continues to move forward and to respond to new discoveries. Alongside this wave of artists working in labs, scientists are adopting practices from outside of the scientific method to bring new insights and ways of thinking into their research. As mentioned, Mark Neyrinck has been using origami to help him gain a new understanding of cosmic structures, increasing knowledge in both directions, from science to origami, and from origami to science; Ansuman Biswas and Chamkaur Ghag are reflecting on how meditation practice might help to develop a more contemplative and potentially embodied science; while Flaviu Cipcigan is interested in how certain forms of dance are distinctive and effective forms of bodily knowledge and wonders how science – currently limited to mental knowledge – would look like as an intimate, embodied knowledge.

Even simply working in proximity to the artists who venture into their domains, scientists can gain from the exchange of ideas and approaches that emerge from these encounters. James Wells, the theoretical physicist who was the scientific partner for CERN's artist-in-residence Julius von Bismarck, talks about the value of having a lengthy exchange with an artist. He found that their rigorous processes were something that they had in common, but that an artist cares much more about the impact of their work on people, whereas "scientists don't generally think that way; they don't understand the impact that their work, and what they choose to do and how they do it, has on students and other scientists and the general science community."[13] One of the things he'd learned from Bismarck was that "it pays to think quite a bit about the impact of one's work; all the way from the grand questions of 'is this right for humans to be doing?' to day-to-day things." He also valued having someone around who sees the world in a different way, whose influence could shake up accepted mindsets, noting that artists start from what they want to do, rather than what they can do: "they have no barriers to thinking about things."

Such processes of dialogue and exchange with artists can also allow scientists to gain confidence in understanding the "performative," the process by which science comes into being through the scientists' acts of imaginative performance of their own scientific methodologies, and then in starting to articulate that. The workshop we organized and publications and exhibitions like these have shown that some scientists are happy to open up in this way. This is not about a whole new way of working, but about giving them "permission," as it were, to place themselves – the whole person – in relationship to their science.

For artists who work with scientists and scientific ideas, theirs is not a process of representation or illustration or interpretation. They are motivated rarely by a desire to educate or inform people about science, but simply by their own fascinations. However, such artists play an important role in the process of cultural assimilation of science by society, of making science intimate. Some of the ways in which contemporary artists help to make fundamental science at the edge of knowledge and at the furthest reaches of scale both personal and intimate are evident in the case of Nahum Mantra, an artist who uses hypnosis in his practice. In his experimental performance text, Nahum probes our potential to have a personal, intimate encounter with the elusive, ephemeral neutrino. Fiona Crisp, in her project *Material Sight,* explores how we might approach the remoteness of fundamental physics by being put in a sensory relation to the "endeavor" of physics. To this end, she has undertaken a series of short residencies at sites of fundamental physics, including the Laboratori Nazionali del Gran Sasso, the world's largest underground research center for particle physics sited at the heart of a mountain in Italy, and Boulby Underground Laboratory, which uses the UK's deepest mine in its search for dark matter. During these visits, she conducted interviews with several physicists. Her resulting installations use non-documentary photography, moving image, and sound to build a tangible sense of encounter with the laboratories and extreme environments where experimental and theoretical physics is performed.

A focus on the unheeded material spaces and overlooked equipment of physics is also at the heart of Semiconductor's film *Magnetic Movie,* in which the artists' visualizations of usually invisible magnetic fields around equipment in a laboratory convey a sense of these normally overlooked lumps of lab equipment as animate and almost sentient. Phil Coy uses poetry and images to give a lyrically different perspective to the equipment that physicists use, while Harry Lawson, through his expanding, reconfiguring collection of rocks, ponders on different relationships between technology, geology, knowledge, and time, what of our culture and knowledge will be preserved, and what may disappear into dust. His musings are reminiscent of an

Arthur C. Clarke story about a machine that sent probes out to the ends of the universe, found the answer, and then realized that there was no one left to tell it to.[14]

Reflecting on the architecture of physics on a geopolitical scale, Blanca Pujals's *A Synthetic Universe* explores how elementary particle physics laboratories have become vast assemblages of people, materials, technologies, and politics, mapping particle colliders, neutrino observatories, and emissions from nuclear test sites and power plants to visualize relationships across the planet. Jol Thomson also begins from the geographies and spaces of physics. From the starting point of his journey to Lake Baikal in Siberia – having been partially transformed by the Russians into a vast neutrino telescope – he reflects on the things that we think we know, but know inaccurately, echoing both Ghag's and Malina's musings on the magnitude of what we don't know about the universe.

From Dewey's Transaction to Barad's Intra-Action

Dewey believed that participating in art (he did not differentiate between fine art and popular art, or between its creation and reception) was an instance of a special type of experience that gives life meaning – that artistic activity enriches our lives by investing everyday actions and objects with new meaning and alters how we perceive the world:

> The function of art has always been to break through the crust of conventionalized and routine consciousness. Common things, a flower, a gleam of moonlight, the song of a bird, not things rare and remote, are means with which the deeper levels of life are touched so that they spring up as desire and thought. This process is art . . . Artists have always been the real purveyors of news, for it is not the outward happening in itself which is new, but the kindling by it of emotion, perception and appreciation.[15]

In 1949, Dewey introduced the term transaction as a philosophical way of understanding the human as an "organism-environment" and of human life as not separated into mind and the world outside it. His transactional perspective makes no radical separation between the subject and the object of knowledge. Knowing in this view is a cooperative, flexible, and open process, and all human behaviors, even the most advanced forms of scientific inquiry, are processes of the whole organism in transaction with its environment.

Throughout the 1920s and 1930s, physicists had struggled with the enigmas and apparent paradoxes presented by developments in theoretical physics,[16] and Dewey himself had been thinking about the implications as early as 1929.[17] He refers to a discussion between the physicists Albert Einstein and Niels Bohr, who took very different positions, remarking of Einstein that: "in contrast with his transactional (i.e. free and open) treatment of physical phenomena, Einstein remained strongly self-actional (i.e. traditionally constrained) in his attitude towards man's activity in scientific enterprise," contrasting this with Bohr's position: "a much freer view of the world that has man as an active component within it, rather than one with man by fixed dogma set over against it."[18]

Einstein and Bohr had been debating the metaphysical issues raised by quantum mechanics, such as wave-particle duality, which had shown that the nature of the observed phenomena changes with the way the experiment is set up. Werner Heisenberg had proposed the "uncertainty principle" to resolve this wave-particle dualism, which focuses on what knowledge we are able to have about a particle's properties. Bohr's proposed solution was more ontological. His principle of complementarity suggests that properties such as momentum and position have no physical reality independent of the observer, and that descriptions such as "wave" and "particle" refer not to independent physical objects, but to different and mutually exclusive phenomena. For Bohr, as for Dewey, we are ourselves part of the reality we are investigating. There is no clear separation between ourselves as investigating subjects and the world as investigated object. They constitute a whole, and the interaction between the observer and the observed is also an inseparable part of the phenomenon. The idea of the experiment being independent of mind or independent of context is an illusion.

The American physicist and feminist scholar Karen Barad takes her inspiration from Bohr to develop her theory of agential realism,[19] however, while Bohr focuses on physics and laboratory experiments, Barad extends his insights of entangled subjects and objects more generally, regarding them not as a particular property of nature, but "the very nature of nature."[20] She has coined the term "intra-action," a key concept of agential realism, to emphasize the mutual participation of subject and object. Some of her ideas link closely to Dewey's notion of transaction via their mutual interest in Bohr.[21]

As with Bohr, Barad considers herself a scientific realist, albeit in a new form. In a similar way to that argued by Ghag and Biswas in their conversation, her view is that, by understanding that the scientist is always part of the apparatus, this can make scientific work more accurate and more rigorous. She believes that this kind of "social construction of science" critique actually makes for better and more credible science.

Forming the Great Community

John Dewey relates the role of art, science, and experience to the goal of achieving a genuine democratic society.[22] If society is to be transformed into a "great community," as he calls it, then the social transmission of knowledge and tools is needed, which requires difficult kinds of inquiries combined with a "subtle, delicate, vivid and responsive art of communication."[23] Art functions as a universal mode of language. Although a specific artwork may not speak to a particular person, art surrounds us and, in its myriad forms, remains at the center of human experience.

The art of science – or of history or politics or any art form – all have the same "material": they are made up of the interactions between the live creature and its environment.[24] None are context- and culture-free. Scientific research responds to specific problems, and thoughtful scientists understand the social significance of their work. Science is not just the accumulation of facts, or of endless measurements. For science to play its full role in society or community, it needs to remain connected to the conditions that inspired it. The danger of scientists failing to consider the moral, social, and technological implications of their research means that science "becomes brutal and mechanical."[25]

Physics – and science in general – needs to realize, and to realize the implications, that science is done by humans for humans. More vitally, it must reconnect with the arts of life, rather than choosing to remain "on an altar in a temple . . . to be approached only with peculiar rites,"[26] where, for many people, it will continue to be regarded merely, and dangerously, as a remote and obscure body of odd belief systems and codes that have little to do with everyday life.

[1] John Dewey, *Art as Experience* (New York: Minton, Balch & Company, 1934).

[2] Kevin J. Pugh and Mark Girod, "Science, Art, and Experience: Constructing a Science Pedagogy From Dewey's Aesthetics," *Journal of Science Teacher Education* 18, no. 1 (2007), pp. 9–27.

[3] This article was originally published in: Fiona Crisp and Nicola Triscott, eds., *The Live Creature and Ethereal Things: Physics in Culture*, (London, Arts Catalyst, 2018)

[4] Broadly, when I use the term "culture," I am referring to the customary beliefs, social forms, and material traits of a particular social group, so when I refer to a culture of physics, I am speaking of the dominant characteristics, beliefs, social behaviors, and traditions of the physics community. When I refer to wider culture, I am talking about the social norms, customs, and behaviors of the society within which physics is situated, which transform over time.

[5] The science wars were a series of intellectual exchanges during the 1990s, primarily in the USA, between scientific realists and postmodernist critics. The scientific realists argued that scientific knowledge is real and accused the postmodernists of having effectively rejected the scientific method and scientific knowledge, while the postmodernists argued that scientific theories are social constructs.

[6] Sharon Traweek, *Beamtimes and Lifetimes* (Boston: Harvard University Press, 1988), p. 162.

[7] Cathrine Hasse, "Cultural Models of Physics: An Analysis of Historical Connections Between Hard Sciences, Humanities and Gender in Physics," in *University Science and Mathematics Education in Transition*, ed. Ole Skovsmose, Paola Valero, and Ole Ravn Christensen (Boston: Springer, 2009).

[8] Ibid., p. 122.

[9] Ibid., p. 123.

[10] Ibid.

[11] Quoted in Nicola Triscott and Fiona Crisp, eds., *The Live Creature and Ethereal Things: Physics in Culture* (London: Arts Catalyst, 2018).

[12] Roger Malina, "Intimate Science: Or Artists in the Dark Universe," a talk given at Sonic Acts (2010), https://vimeo.com/12920830 (all URLs accessed in January 2019).

[13] CERN, "Interview with James Wells (Collide@CERN) for Swiss Info" (2012), https://videos.cern.ch/record/1483712.

[14] Ansuman Biswas relayed this story in his conversation with Chamkaur Ghag (although this is not in the edited version included in this book).

[15] John Dewey, *The Public and Its Problems* (New York: H. Holt and Company, 1927).

[16] Dewey learned about the breaking developments in quantum mechanics in the late 1920s firsthand from his daughter Jane, who had conducted postdoctoral research with Niels Bohr, Werner Heisenberg, Erwin Schrödinger, and other physicists in Copenhagen.

[17] When Dewey wrote *The Quest for Certainty.*

[18] John Dewey and Arthur F. Bentley, *Knowing and the Known* (1949; repr., Boston: Beacon Press, 1960), p. 108.

[19] Karen Barad, *Meeting the Universe Halfway: Quantum Physics and the Entanglement of Matter and Meaning* (Durham NC: Duke University Press, 2007).

[20] Ibid., p. 6.

[21] Matz Hammarström, "On the Concepts of Transaction and Intra-action," The Third Nordic Pragmatism Conference, Uppsala, June 1-2, 2010, https://internt.ht.lu.se/media/documents/persons/MatzHammarstrom/On_the_Concepts_of_Transaction_and_Intra_action.pdf.

[22] John Dewey, *Experience and Nature* (London: George Allen & Unwin, 1929).

[23] Ibid., p. 184.

[24] Dewey, *Experience and Nature.*

[25] Ibid., p. 382.

[26] Ibid., p. 381.

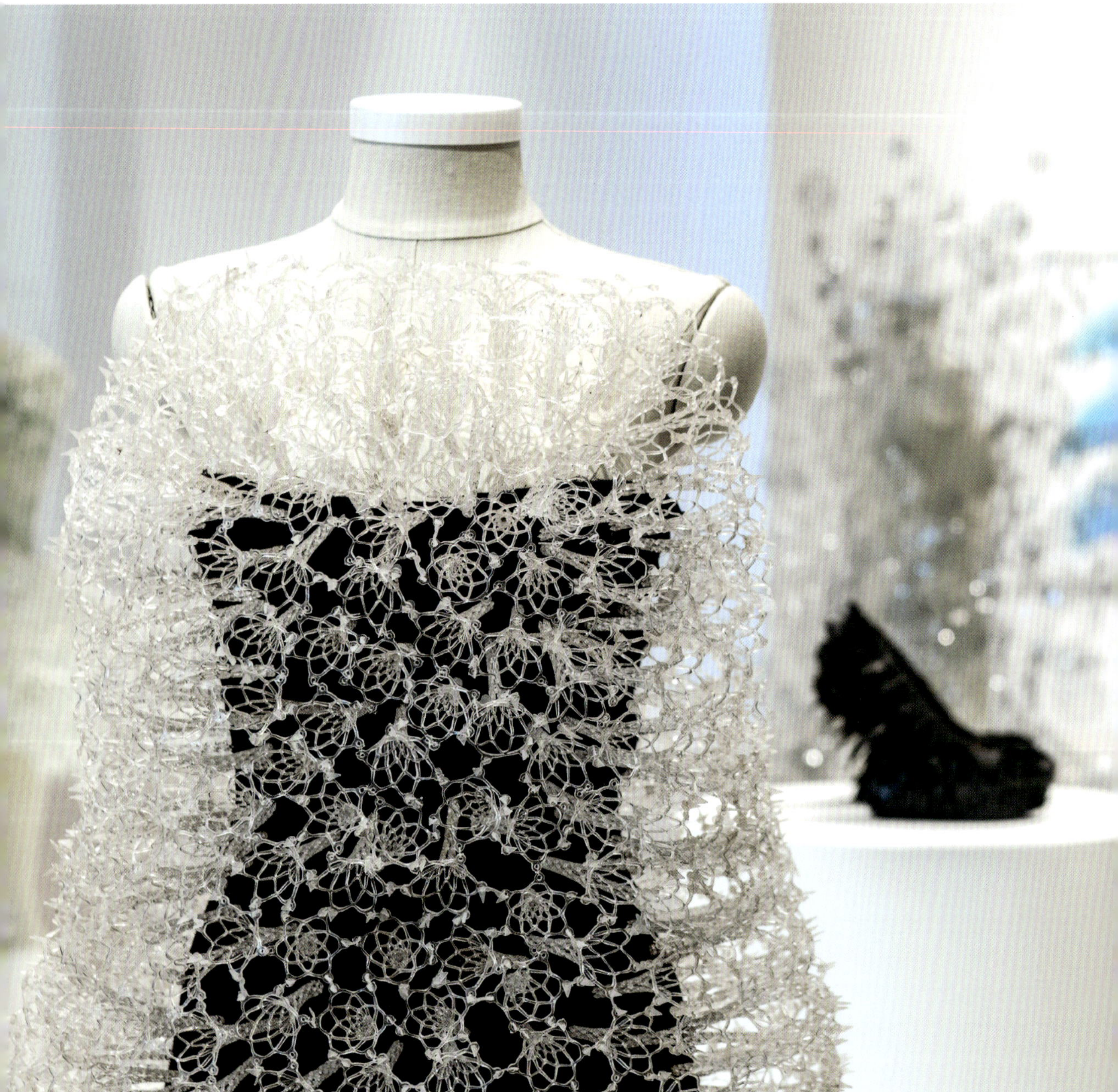

Dress from the *Magnetic Motion* Spring/Summer collection 2015
Iris van Herpen

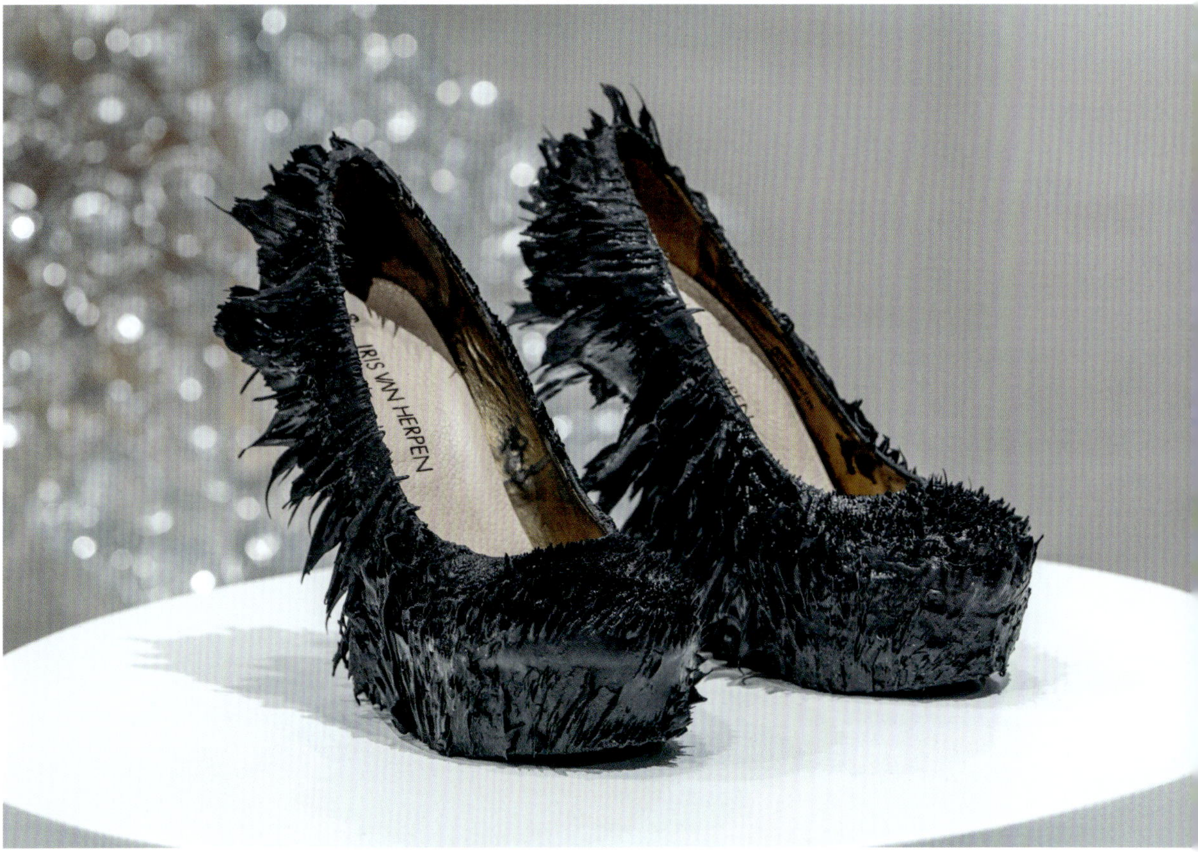

Shoes from the *Magnetic Motion* Spring/Summer collection 2015
Iris van Herpen

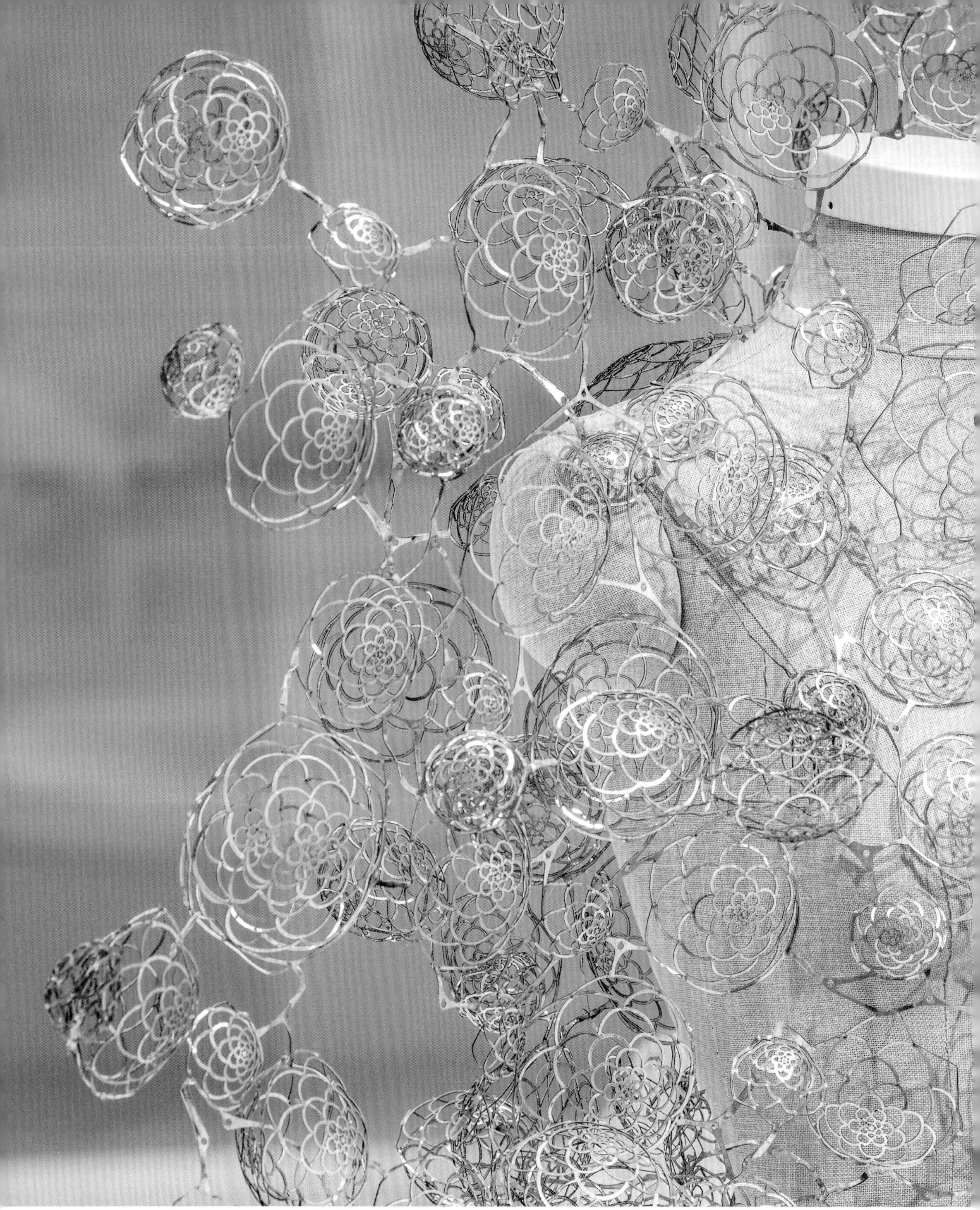

Detail, Dress from the *Aeriform* Spring/Summer collection 2017
Iris van Herpen

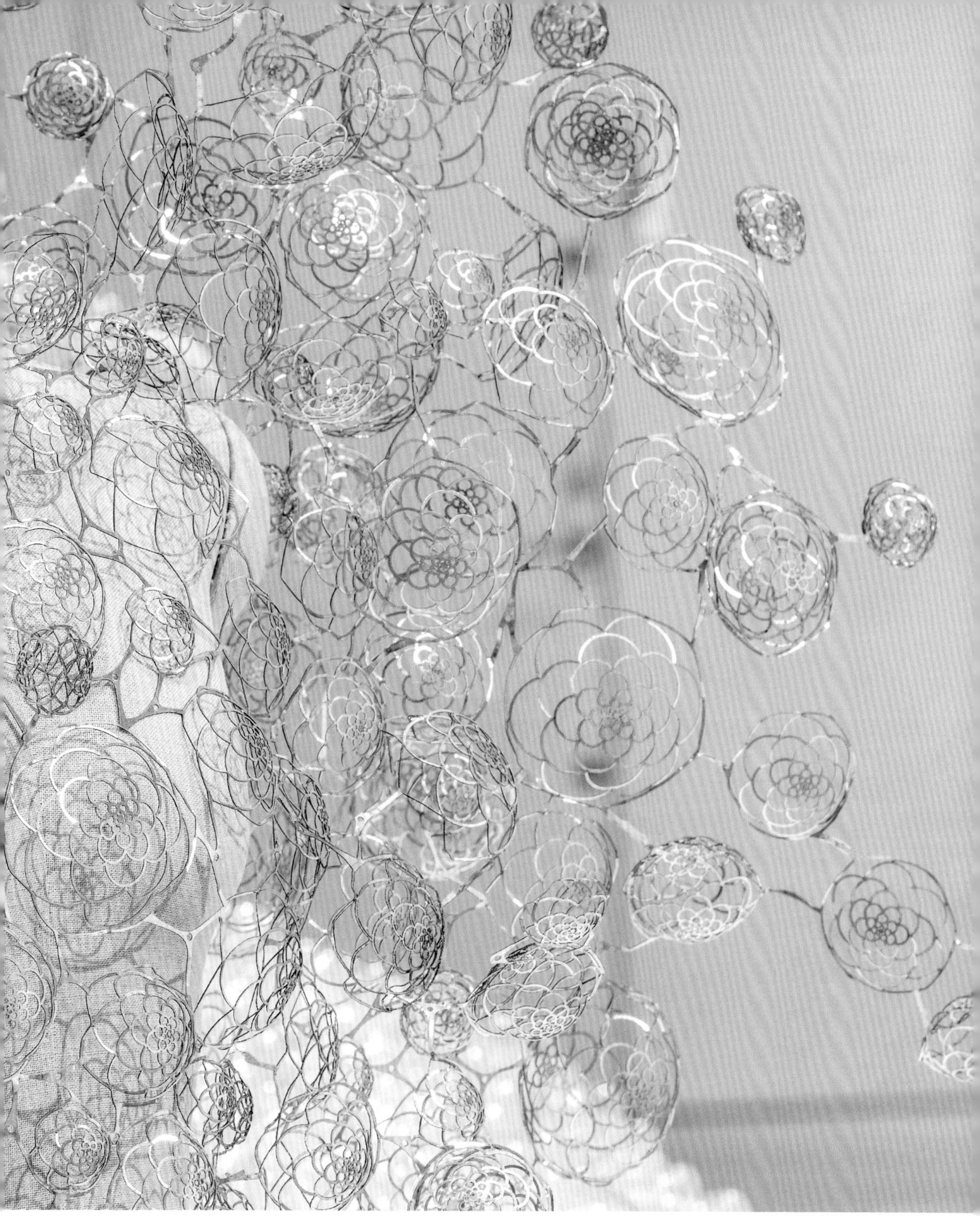

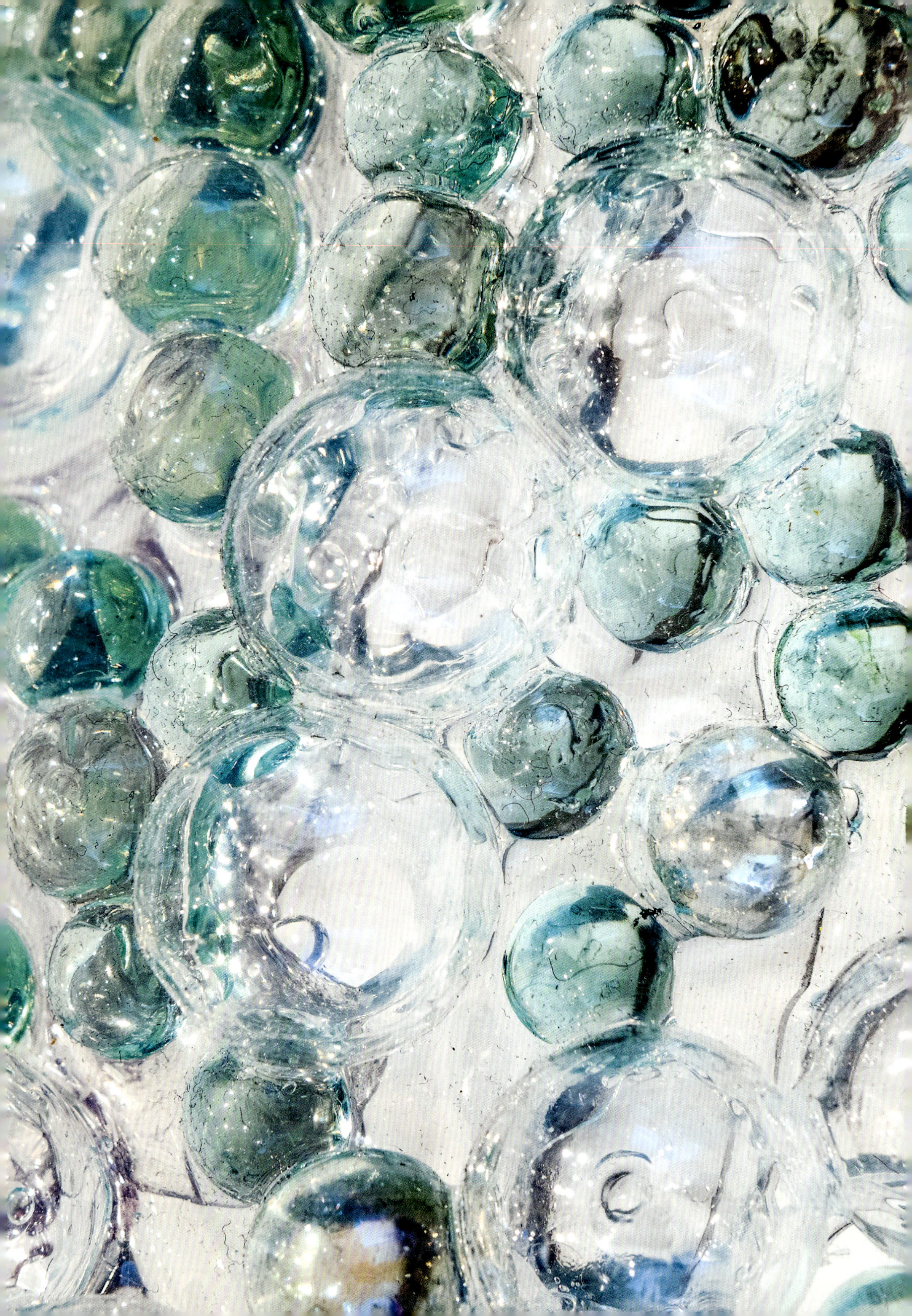

DIPTYCHS

Ariane Koek

Matter Time Space Gravity Light Entropy

One artist
One physicist
One natural world phenomenon
Two ways of seeing and experiencing the world.
These are fragments of 12 conversations
About 6 different phenomena
Recorded separately across time and space
Only meeting here
Between the pages of this book

By reading the reflections of the artists and the scientists, the reader discovers connections and the differences in their ways of looking at the world.
It's in this space
that the imagination grows.

Detail
Bubble dress from the *Sejiaku* Autumn/Winter collection 2016
Iris van Herpen

ON MATTER:

Iris van Herpen

Water, transformation, concentration, nature, the beginning, energy, my body, creation, expansion, seduction, knowledge . . .

Matter is creation, it's evolution, it's nature, it's us. It's the source of all energy and the source of all of our questions. I am fascinated by matter in all her shapes and all her mysteries. Perhaps I am even in love looking around me, at how it continuously transforms in front of my eyes. Seeing water move, for example, makes me feel useless as an artist as nature is the most ingenious artist itself. But at the same time, matter and metamorphosis are my muse. I look for beauty at the intersections, the intersections between art and science, between precision and chaos, between the artificial and the organic.

Personally, matter to me means misbehavior. I am seduced by the behavior and the misbehavior between the body and her architecture, her shelter. I think it's the interaction between the two that makes a challenge: sometimes a garment is completely in tune with the body, together they dance. Sometimes it's the contrary: the garment influences the body and this, too, can be very inspirational ground – it challenges me to think beyond today's definitions of a garment. What will the unexplored fields of human-designed biology bring us? How will evolution shape our bodies to come? Will our own matter become transformable? Will we become able to instantly change our skin, color, texture, and identity, expressing our thoughts and emotions like a living painting? I can be very jealous of the octopus . . .

Matter is both my tool and my muse, but sometimes also my enemy, as my mind always moves faster than today's reality. Matter is all and the sculptor of life. In my process, I find it important that my materials are guiding me more than my mind. Ideas come from memory, and they therefore feel more predictable and momentary than the more random and unpredictable processes that arise without thoughts between my hands working with the material. There is lots of matter floating in between us; perhaps there is knowledge floating in between people that is yet to be discovered, undefined, like a gut feeling. Going deeper and deeper into the work as an artist, it is tempting to let ideas and concepts define the design; it's more controllable, but to me this feels too much connected to my past. I force myself to let my materials guide me intuitively.

My creative process is exploration, experimentation, and transformation. Mostly very simple: a material inspires me to start designing. This can be as pure as being a piece of leather, or as invincible and challengeable as water.

I start with getting to know the material, finding its language through its strengths and weaknesses. Then after seeing a first test, a new idea comes and the adjustment process starts that can go forward and backward hundreds of times. Along that intense path of trial and error, all logic becomes chaos. That moment of surrender to the irrational is where to me the best designs come out, where intuition and emotion are talking and are at their most powerful. That is the only way for me to find a natural beauty that can touch that deeper layer, having the power to change me emotionally. I really believe that wearing something exceptionally beautiful, made with lots of thought and lots of care, is like listening to a beautiful piece of music that overwhelms you, giving you the power to communicate with people around you on a deeper level, an unconscious level, like sound or smell can do.

Art is linked to our deepest desires, moods, and our most personal expressions. The ability to transform a material feels like magic, like an alchemist's dream. Even a very small transformation – like the pleating and stitching of a fabric into a three-dimensional texture that floats together with the body like a jellyfish moving underwater – can impact me. It's not so much the outcome; it's more the process of metamorphosis that strangely enough makes me feel more a part of and more aware of this planet.

ON MATTER:

Michael Doser

Matter is massive, compact, heavy, light, transparent, flimsy, an illusion.

Matter is the building block of everything. The particles that make up matter are what make up everything in the universe or at least everything that we know and can probe. The interactions of matter with fields are what allow us to see the effect of the matter, of the particles, through the different forces, through light, through other interactions, through gravity. We can probe the structure of the universe in this way. In a sense, matter is the scaffolding that allows us to study the structure of the universe.

Without matter, nothing is possible. No complexity is possible. You end up with the universe – I can contrast this maybe with a universe that would consist only of light, where there is no interaction, there is no matter. This is an exceedingly boring place, nothing happens, no planets, no stars, no galaxies, no life, no thinking, no intelligence.

What gives the sculpture life or form is a specific combination, a specific choice of how this matter is combined, how it is put together, how it is structured. The really interesting aspect is then which particular combination the matter takes in order to form, on the one hand, a butterfly and, on the other, a tadpole.

The current picture is that everything is made of one of six types of quarks or one of six types of leptons, electrons, neutrinos, heavier electrons, heavier neutrinos. These twelve particles are the building blocks of everything. They're sort of the bricks, which are combined among each other through forces: the electromagnetic force, the strong interaction which holds everything together at the heart, the nucleus of an atom, is only stable because of the extremely strong gluing power of the strong interaction, and then gravity.

From a physicist's point of view, that model explains everything there is to know about the universe.

If you look at the overall picture of the universe, however, ordinary matter is really a minority compared to the other components of the universe, but it's the only part that we can actually probe because it's the only part that interacts with light so that we can see it. It's the only part that is really understood.

To me, as a physicist, matter means many things . . . Matter plays an extremely important role in my life because it gives me a wall to bang my head against so that I can try to solve the questions that I'm working so hard to come to grips with.

My job is to try to test our understanding of what matter is, what antimatter is, and to see if we can find any unexpected surprises by measuring their properties extremely precisely. If there were no antimatter, then the universe would be empty because there would not be any matter either.

The existence of antimatter is a precondition to the existence of matter, and thus of the universe in which we live. But at the same time, the existence of antimatter is the worst enemy of the existence of matter if the two could meet. If you go back to the moment of the Big Bang, the whole energy of the Big Bang produced pairs of particles and antiparticles in incredible quantities.

But if that were the whole story, then these pairs of particles and antiparticles could meet up again, destroy each other, annihilate and transform back into energy and ultimately into light, and you would end up with an empty universe consisting only of light. That's not what happened. What must have happened – and this is the best model that we have to date – is that for every ten billion (10,000,000,000) particles of antimatter, ten billion and one (10,000,000,001) particles of matter were produced.

In interacting with each other, the ten billion particles of antimatter destroyed themselves with ten billion particles of matter, and one particle was left over out of these ten billion and one (10,000,000,001). It is really amazing that everything that we experience – the matter we feel around us, the buildings, the chairs, the food we eat, the air we breathe – that all this is the leftover, the tiny little remnant from this initial cataclysm, this initial battle.

What I'm trying to do, what I'm spending my life on, is trying to find out if there's some slight discrepancy, slight difference – minuscule, microscopic – between matter and antimatter that could explain why the universe is the way it is and not empty.

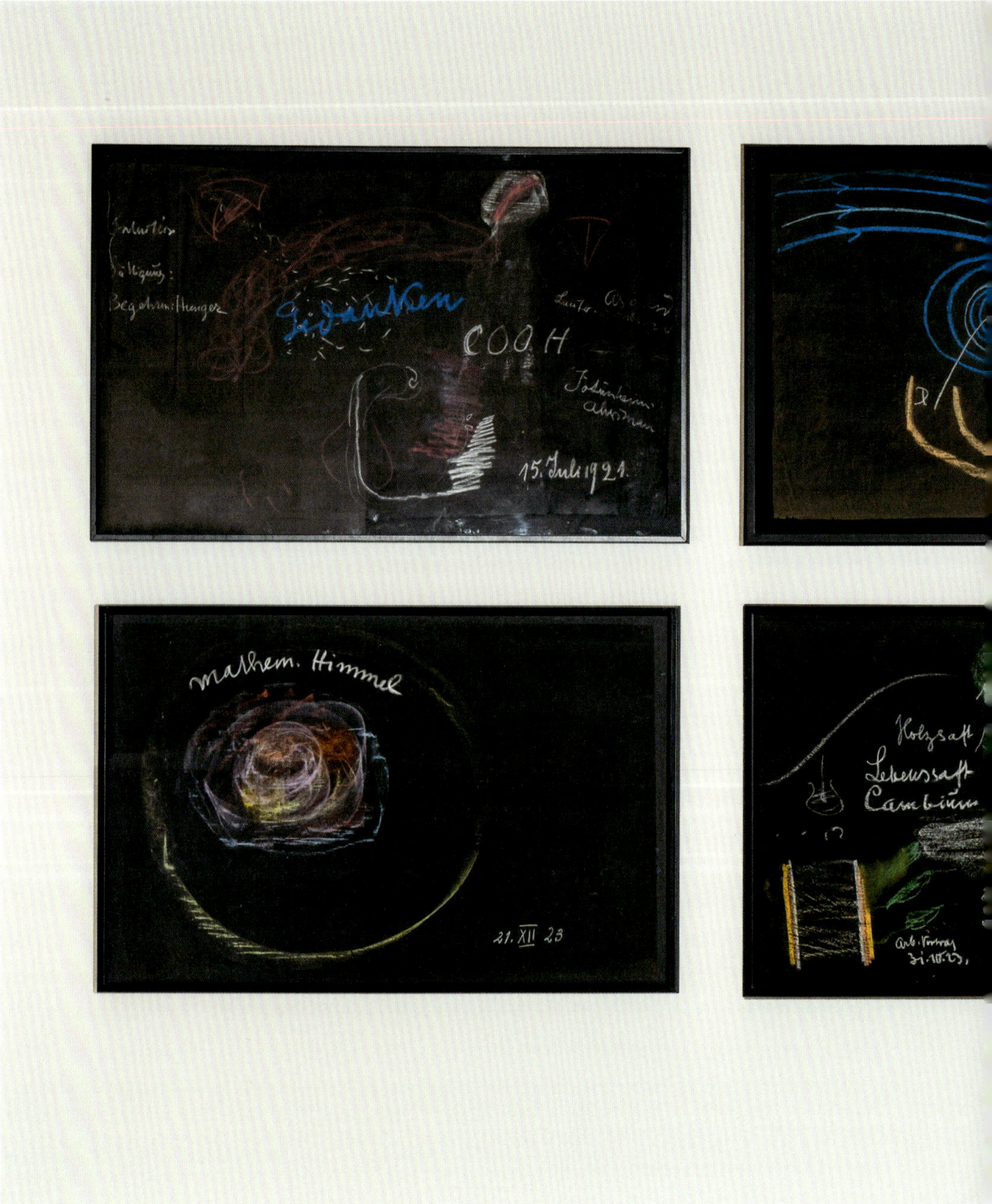

Detail from *Untitled*, installation by Goshka Macuga, 2018
Blackboard drawings from various lectures
Rudolf Steiner
1921–1924

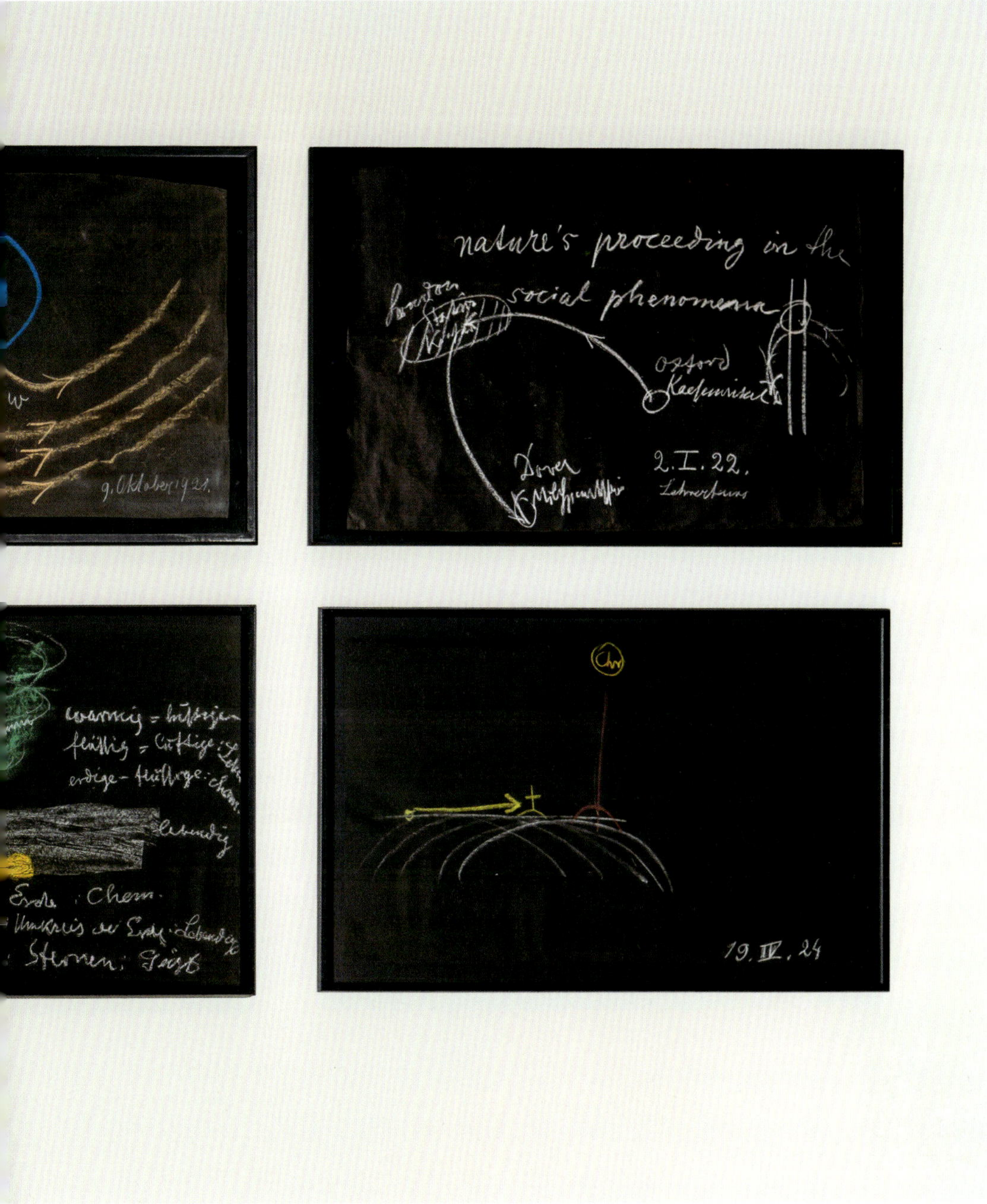
9. Oktober 1921.
nature's proceeding in the
social phenomena
London
Oxford
Dover
2. I. 22.
Erde : Chem.
Sternen: Geist
19. IV. 24

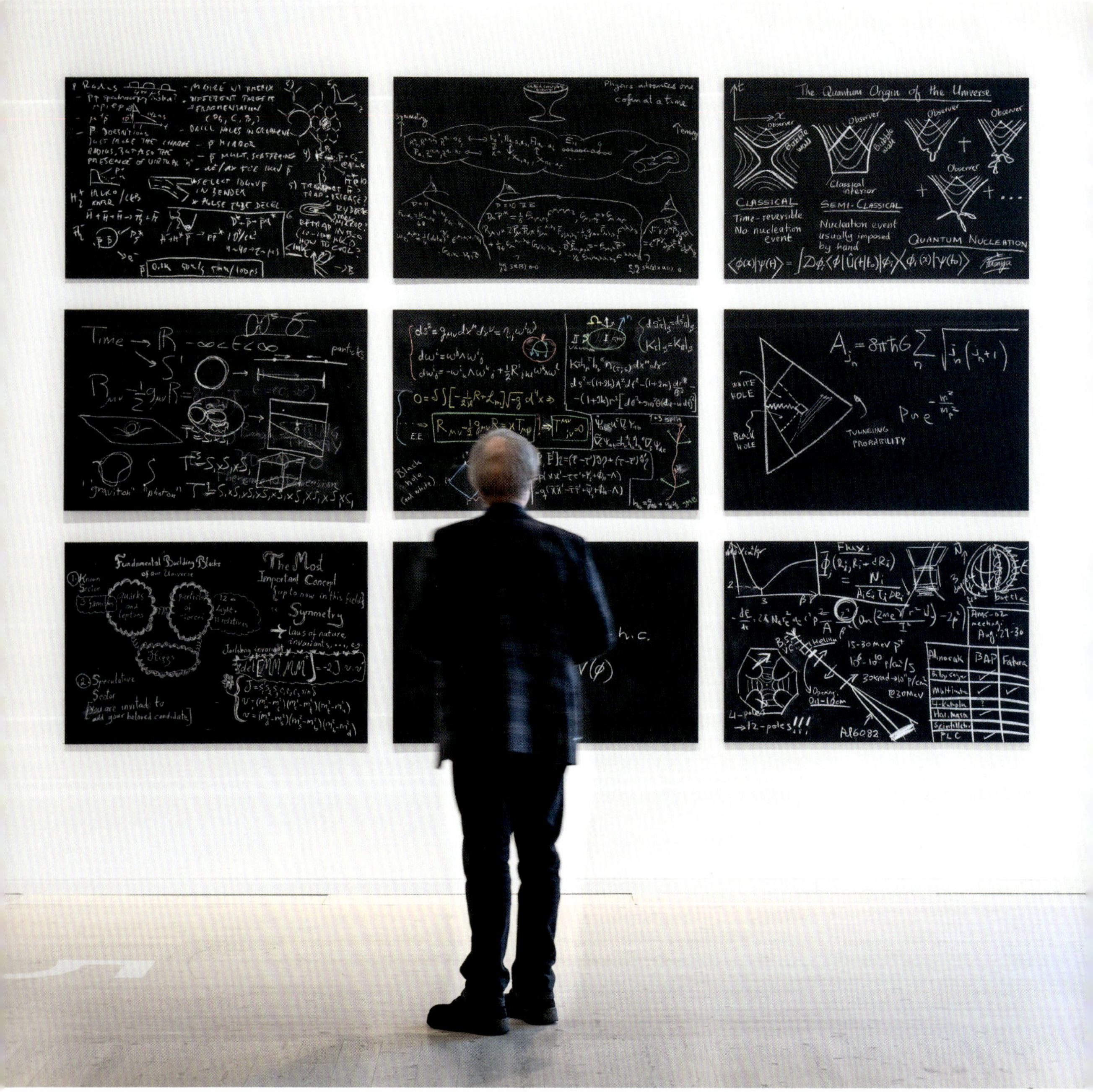

Detail from *Untitled*, installation by Goshka Macuga, 2018
Blackboards written with equations by contemporary physicists
Top, left to right: Michael Doser, Peter West, Hiranya Peiris
Middle, left to right: Subodh Patil, Michael Bradley, Carlo Rovelli
Bottom, left to right: Cecilia Jarlskog, John Ellis, Bilge Demirköz

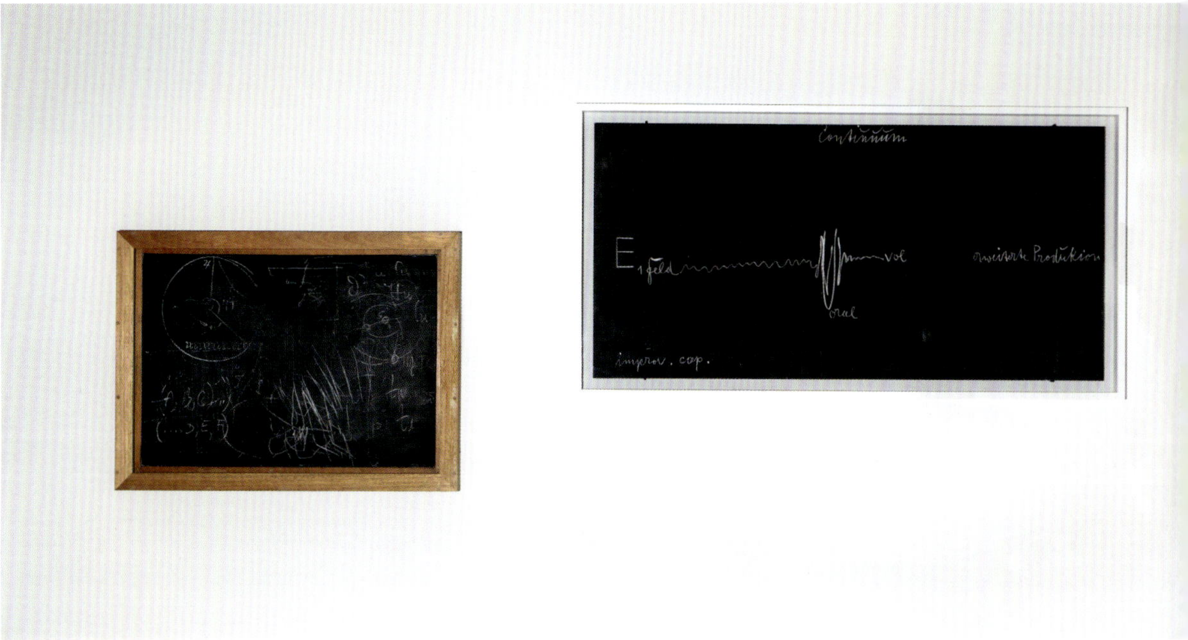

Detail from *Untitled*, installation by Goshka Macuga, 2018

LEFT: Blackboard with white chalk traces
Albert Einstein
1931

RIGHT: *Continum*
Joseph Beuys
1978/1984

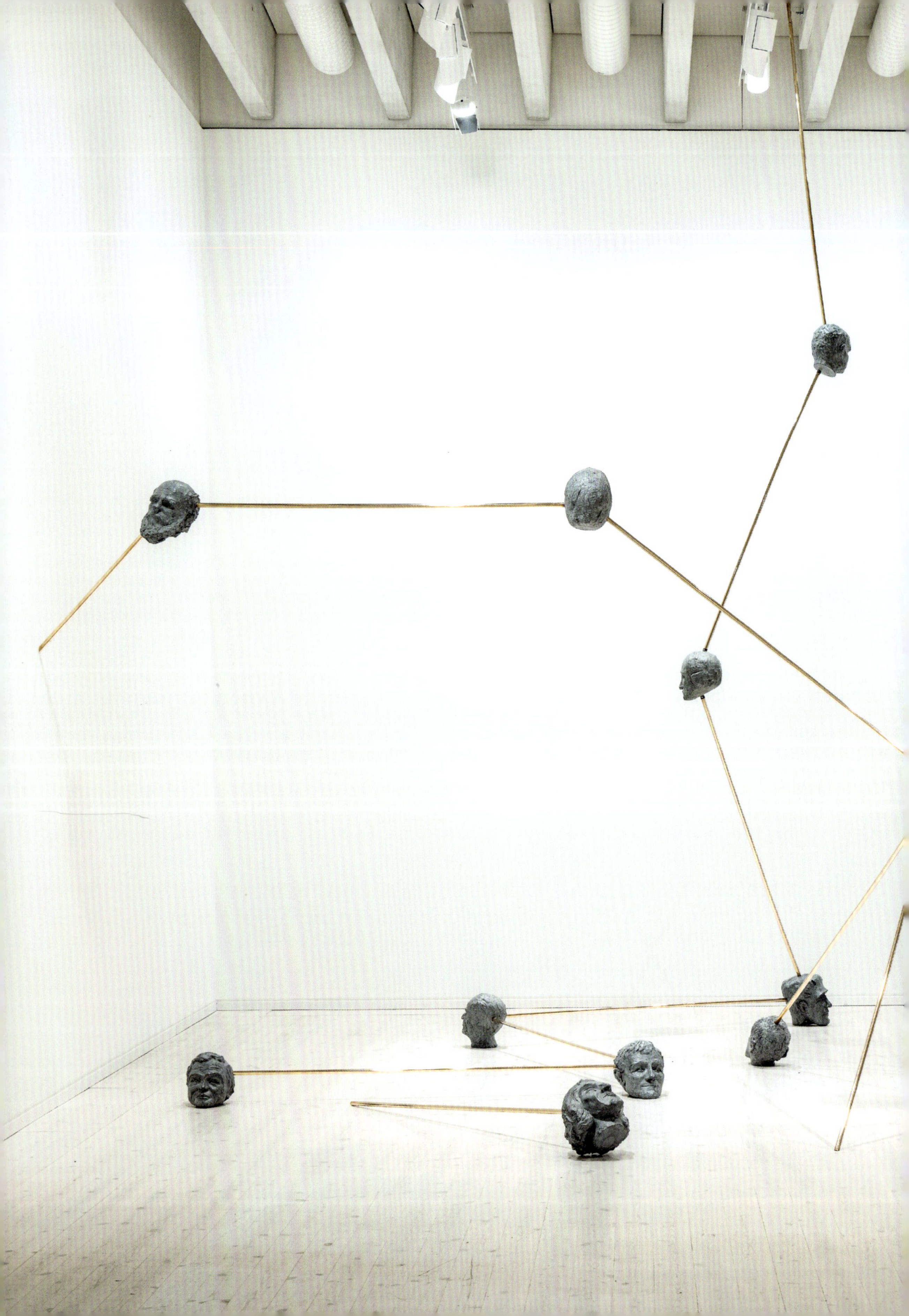

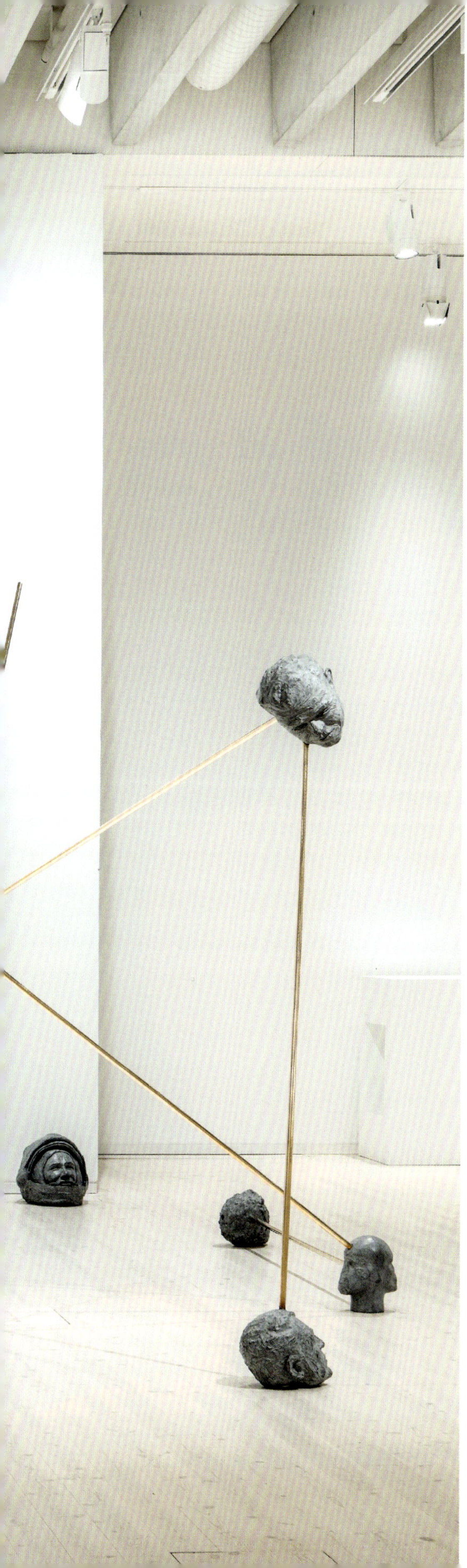

International Institute of Intellectual Cooperation, Configuration 5 & 7
Goshka Macuga
2015–2016

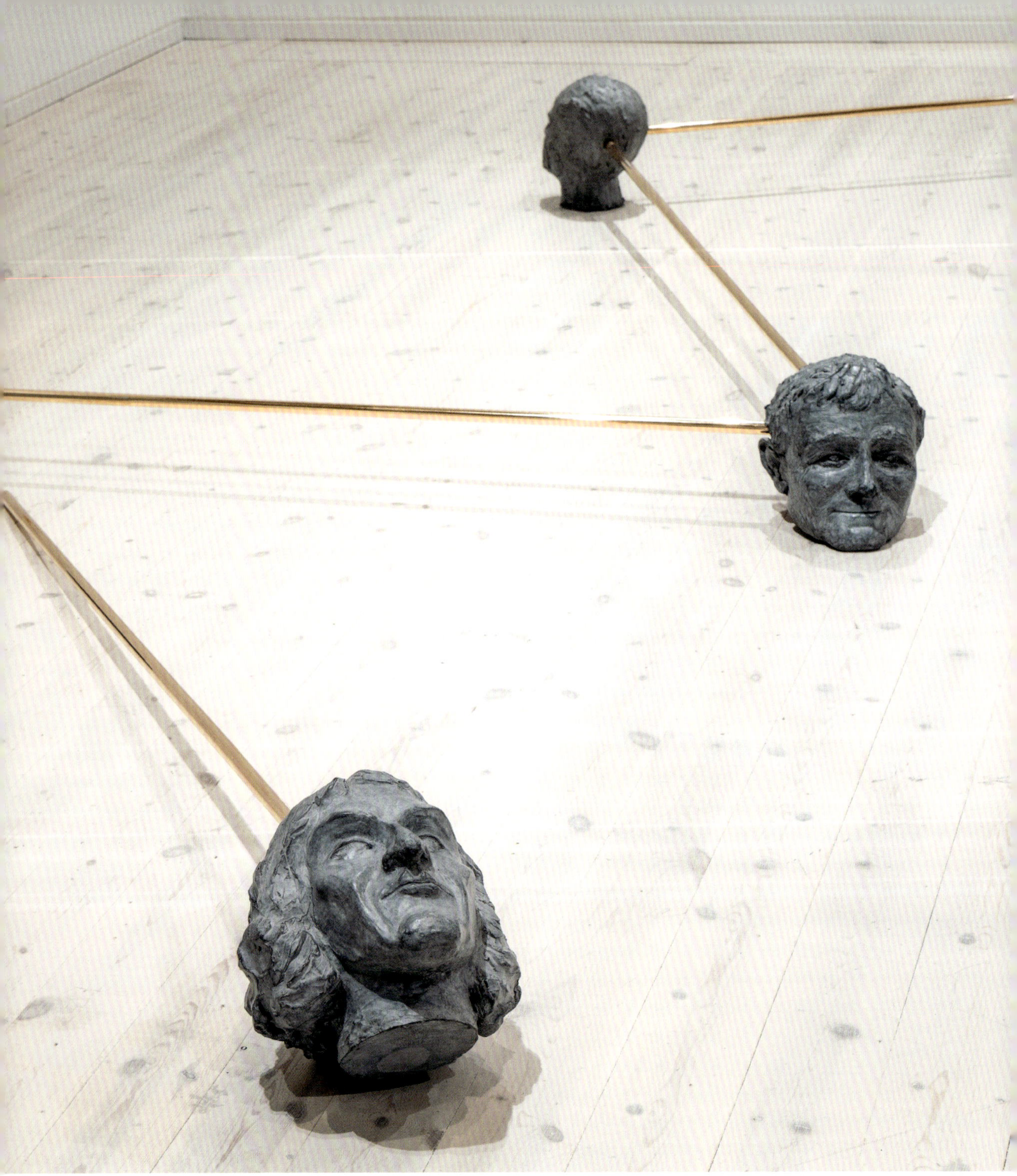

Detail
International Institute of
Intellectual Cooperation,
Configuration 5 & 7
Goshka Macuga
2015–2016

ON ENTROPY:

Goshka Macuga

Disorder, Heat Death

Entropy according to the dictionaries is a measure of disorder in a closed but changing system. In this system energy can only be transferred in one direction from an order to a disorder. Entropy comes from a principle of thermodynamics and sadly sees the future of the universe driving toward a "total disorder" called the heat death of the universe. In 2017, I invited the physicist Subodh Patil to do a presentation about entropy during my exhibition at Witte de With in Rotterdam. I met Subodh at CERN a few years ago, and it was through him that I learned about heat death.

The opposite of entropy would be a measure of a system that can transform from disorder to order. I guess this progression is something that I feel more comfortable with. It is something that I aspire more toward in my everyday life. "Disorder" as such conceptually found a place in my practice in recent years and instantly attached itself to what can be considered a constructive process of reordering or reconstructing or decontextualizing or even renaming.

Paradoxically, the opposite of entropy is information order. Information is something that drives my work and my life.

There are many ways that I can relate to entropy: one is definitely conceptual, therefore more attached to its origins in physics and math. I have been fascinated with the development and use of our scientific language that attempts to describe the universe and its evolution. I'm also drawn to the aspect of entropy that involves the observation of time and change as well as the relationships within. Scientific terminology can often be misleading for people such as me who stopped learning math and physics in their late teens.

But in a looser sense, the term and idea of entropy in art-making can make for an interesting proposition – something like destruction as a process. In 2017, I collaborated with Ahmet Öğüt to create an exhibition for a space in Rotterdam at Witte de With (mentioned previously). For the second part of the show that opened in

September 2017, I interfered with Ahmet's works as well as mine. Engaging destruction as a means to challenge institutional structures, I wanted to question: How far can destruction work to critique, protest, and confront the present socioeconomic, political predicament? In the face of the recent resurgence of right-wing, populist, and nationalistic agendas that have come to dominate the current political landscape, what can be gained by enlisting destruction for social critique, but also anarchic, pointless destruction that reflects our reality, destruction for the pure pleasure of it?

Destruction, chaos, and anarchy can be a driving force in creative processes such as art, literature, fashion, or music. Occurrences of these themes can be found not just in the twentieth century, but also far earlier. The earliest records of human knowledge registered in manuscripts (most commonly religious documents) use destruction and chaos as motifs to describe a dynamic or a consequence of a process we engage within. These early ideas were inflicted on us and used as a form of social control. They're still used in the same way today.

The fear of total destruction or the idea of "the end," especially the end of the world (or even the concept of heat death of the universe) can activate a number of different social responses, from uncontrolled panic and fear to an attempt to revolutionize the existing order to complete indulgence and play. If god does not exist, if salvation becomes meaningless, then heat death will happen regardless of what humanity does – it's indifferent to our existence, so why bother? There was an Austrian physicist and philosopher called Ludwig Boltzmann who tried to define entropy and its connection to probability – he committed suicide. I am not sure if this had to do with the nature of his work and a realization of a certain type of futility of existence, or if it was his mental condition.

We live in a world that we have created with ideas of order. Homo sapiens love patterns; we find them in everything and use them to create structure around us. The design of the cities we live in and our personal spaces focus on the maintenance of order. Our languages, our legal systems, our attitudes toward each other are in continuous negotiation of both positive and negative aspects of order and disorder. If the earth is heading toward disorder, then we ourselves have a big engagement in this process. Entropy doesn't exist without a human perspective, and humanity seems to be conditioned by its destructive nature.

ON ENTROPY:

Subodh Patil

Entropy is one of the most foundational concepts in physics because it underlies why matter and how matter behaves.

Entropy is the number of ways a system can be with a given energy. If I have a ball placed on a floor in the middle of the room, then it's in its ground state. It is not going to fall any further. If I put it up on a table, then it will fall off that table and give a bit of a loud thump. It had some energy, but when I put the ball on the floor, it's in its ground state. It cannot fall any further.

Entropy is the number of ways that a ball can be on the floor. I can put it in the center of the room. I could put it in a corner of the room. I can move it somewhere else. If I count up all the different places I could put the ball, that number is related to the entropy of the system. It counts the number of ways a system could look indistinguishable as far as its energy is concerned.

I think that we all become conscious of entropy quite early on because we have parents who yell at us and tell us to tidy up our room. Things get messed around, things get moved around. When you're told to clean up, make your bed, tidy up your room, what you're doing is you're locally reducing your entropy. Entropy is a way of counting the disorder of the system. There are many, many ways to mess up a room, but there are only one or two ways to keep it quite ordered. Intuitively, entropy is a concept that's not very far away from our daily experience of order and disorder.

Entropy is the only number that we can write down that seems to always go up in time. It seems to actually be the origin of the "arrow" of time. Because we exist in a universe, we are participants in a universe, we are made up of the same stuff that exists in the universe, then the way we are made up causes us to perceive time in a particular way. That's a puzzle because according to physics, the fundamental laws of physics work the same, moving forward in time as opposed to backward in time.

The only description, the only understanding we have of why time has a forward direction – time's arrow – has to do with entropy. As a cosmologist, that's what you're trying to explain. You're trying to explain why the universe is the way it is, how it became the way it is, and where it is going. Where is it going to end? Where we think it's going to end is ultimately in the so-called heat death of the universe. When the universe ultimately attains its maximal possible entropy, everything will be maximally disordered and it will not be possible for anything to ever happen again. That's the ultimate fate of our universe.

To ask what the universe would be like without entropy would be to say that the universe would be in a state of zero entropy, which would mean that its ground state is unique, and that's entirely possible. We don't know if that's true or not, but we're certainly trying to find out.

When I was young, I used to be the kind of kid who would always try to break things open to find out what's inside them to see what makes them work. I was never really ever content with answers to questions that I would ask my parents or my teachers, because they'd always end up with the obvious question, "Why that answer?"

What I found was that whenever I'd ask a question about something – "Why is the sky blue?" or "Why does the salt taste this way?" – then I'd get an explanation, and that explanation had an answer, but then you could ask, "Well, why that answer?" And it just seemed to me that when we kept going, the buck stopped with physics. Physics is the only subject that doesn't run away and look for other things to explain. It has to explain itself, and that really appealed to me. It seemed to be fundamental.

Entropy is not something we control. Entropy is something that's out there, and it's always increasing. It would be like asking, "How can we keep time under control?" We cannot. It's just having its way with us all the time. It's the same thing with entropy.

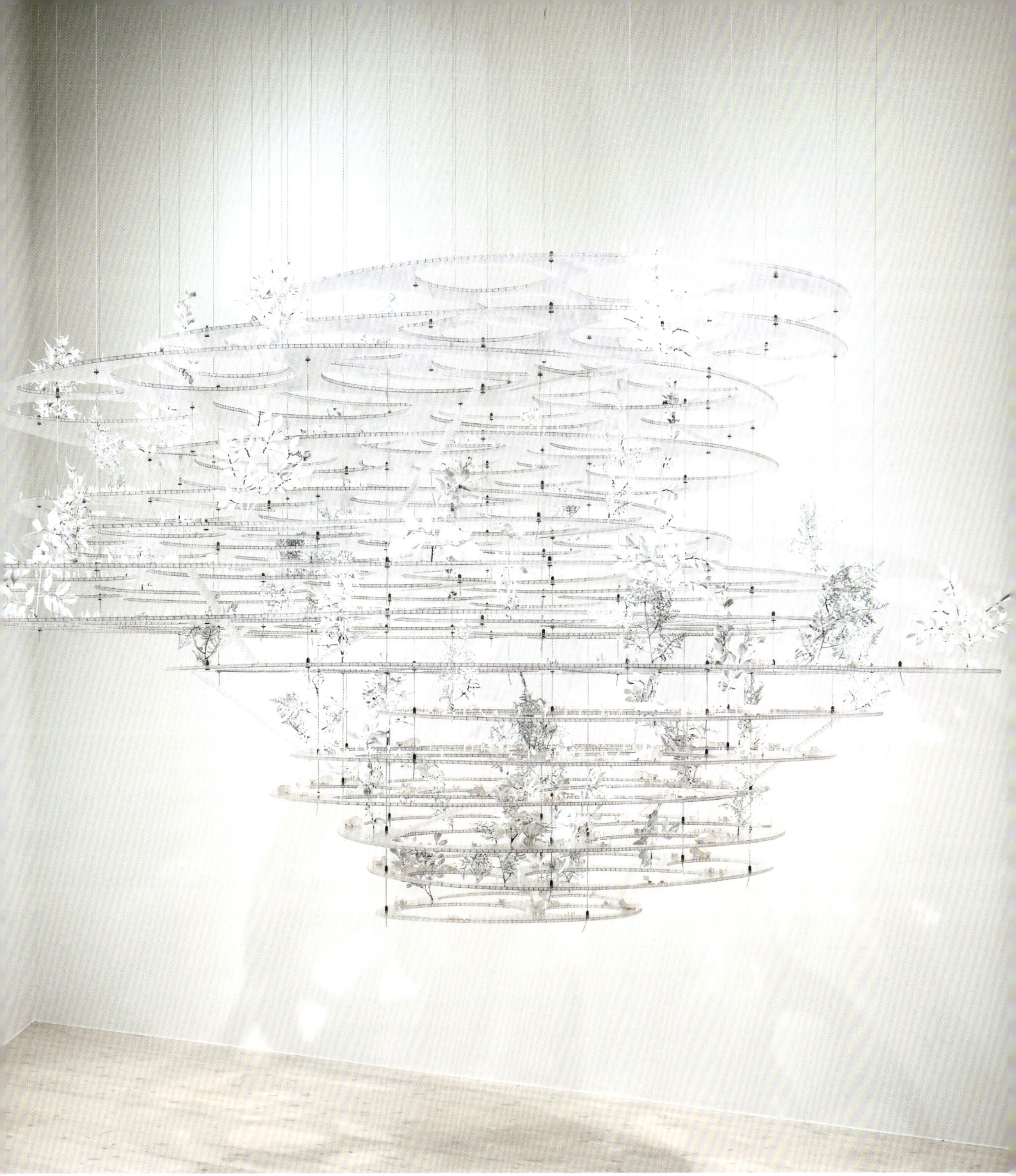

Energy Forest
Sou Fujimoto
2016

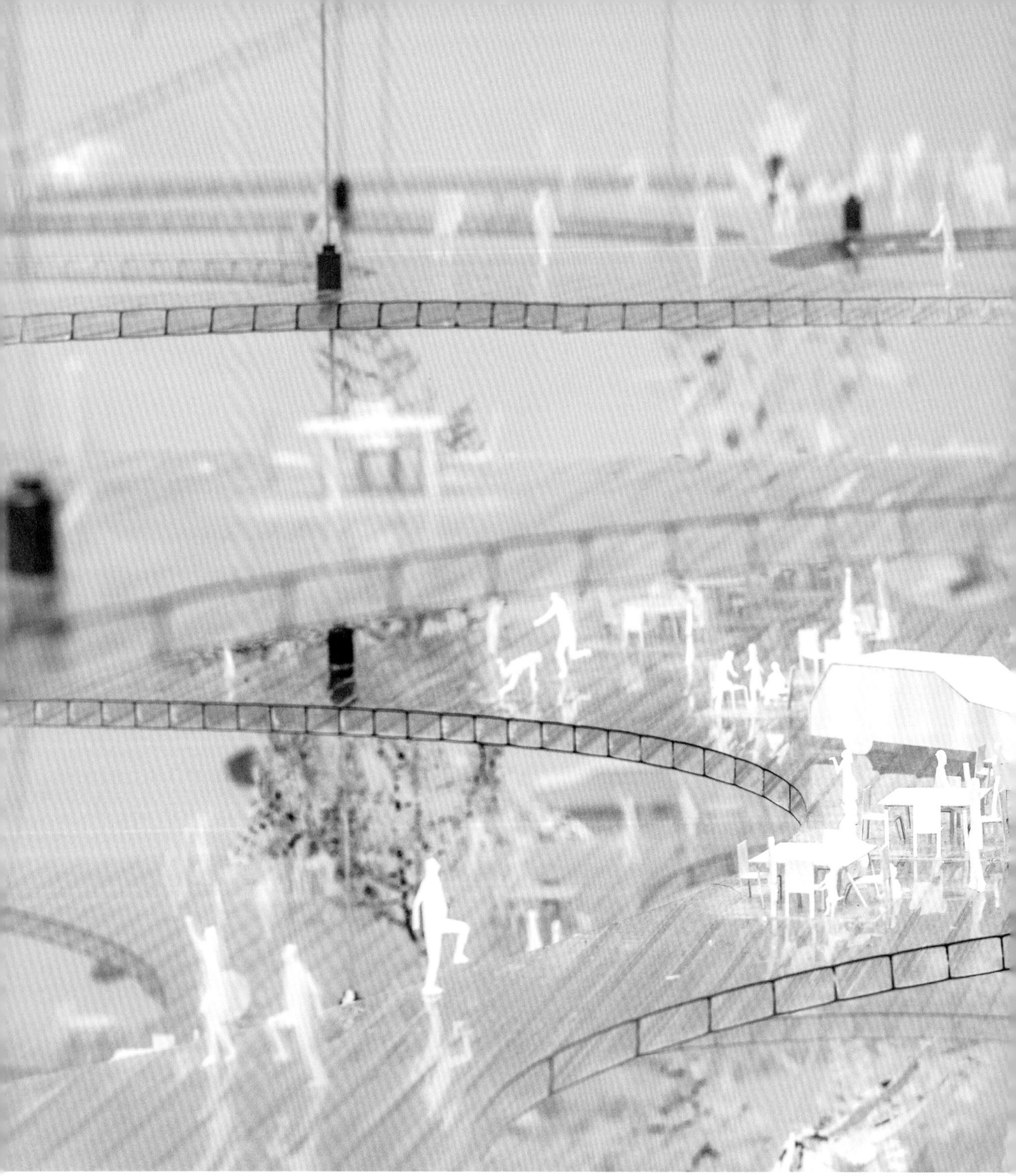

Detail
Energy Forest
Sou Fujimoto
2016

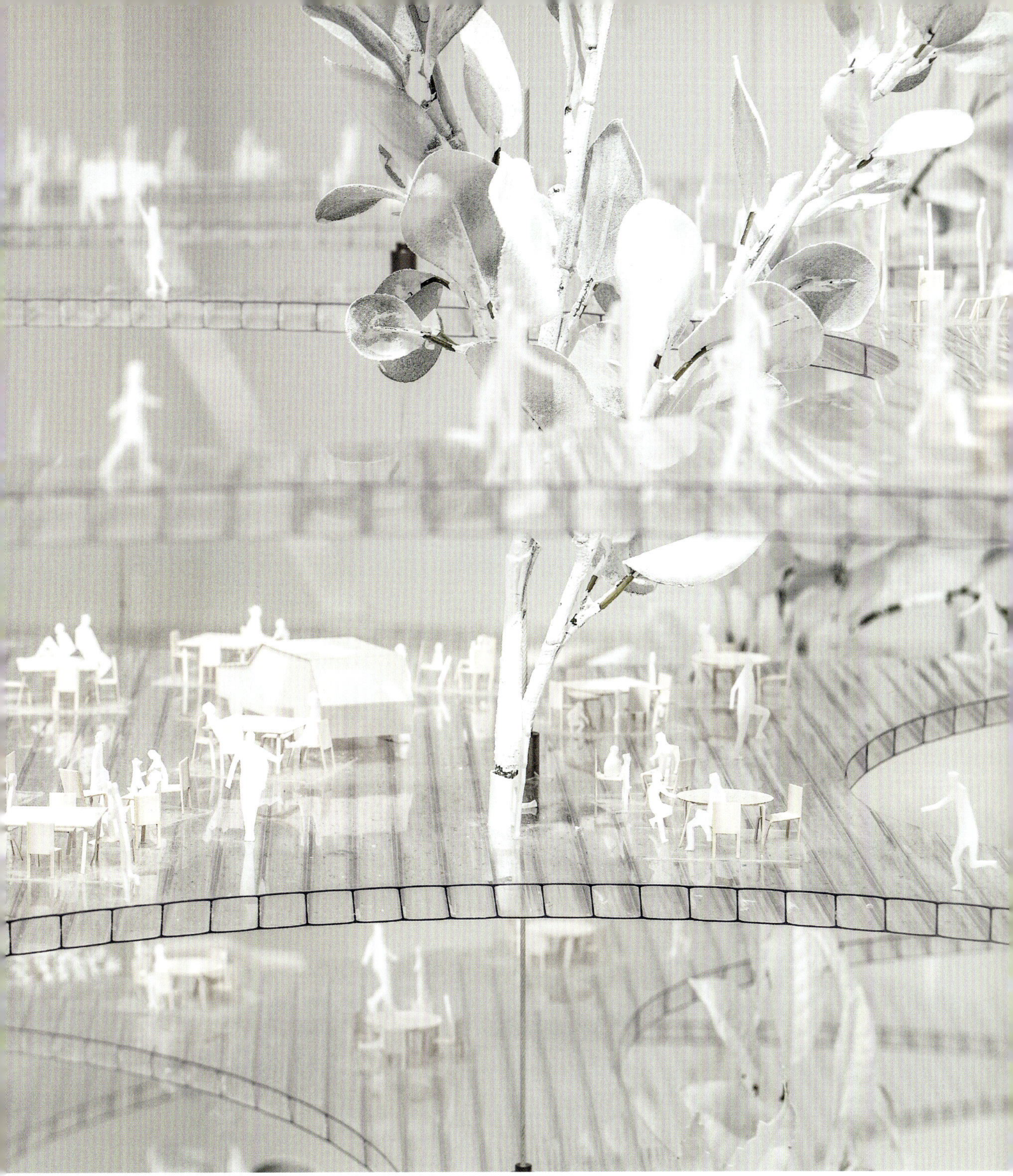

ON SPACE:

Sou Fujimoto

Space, for me as an architect, is like relationships – the relationships between people, relationships between inside and outside, relationships between human life and nature. Because in architecture, of course, we make space, but space is not like the object we just see as an art piece, but the people will be inside to make their own life. The relationship itself is the space.

I myself, before I studied architecture, was really fascinated by physics and the theory of Einstein and the theory of many of those new ways of trying to understand the whole world and the relationships of space and time and so on. The word "space" was originally, for me, from physics, not from architecture.

For example, for Isaac Newton, a long time ago, space was more like a neutral background in a sense, but for Einstein space was more relative, making relationships and changes situation by situation. When I started to do architecture, the architectural space was not just making a huge background of everything, but every time your experience related to each different situation and space, it became more dynamic. I got really big influences from that.

I have a photograph of Einstein in my studio, and recently I moved to a new place, and I brought that with me, because Einstein has been the origin of my thinking. All of this passion to create something new, in a different, deeper way, started from the passion for Einstein.

I was enjoying the space in the forest when I was a child, but, of course, at that time I didn't even know the word "space" – I just felt it. After learning architecture, of course, I learned space, but then, for example, the small winding streets in Tokyo made me feel space. It reminds me of the experiences in the childhood days in the forest. Both are connected to each other – the street, the forest. You are in the middle of many small things surrounding you to protect you, but still you can be free to walk around in the street or forest.

In Tokyo, you are also surrounded by many ugly but many different types of scales and pieces, so you feel rather like you are protected, and still you can be free to walk around in the city. That was the time I realized, "Wow, this is something like space, and it is relating to my childhood memory."

Space is not enemy, not friend, but a part of myself. Then gradually it started to part from myself, something between myself and not myself.

I'm really curious about the perception of future space and architecture. Recently, I got into a discussion with my friend, a specialist in artificial intelligence and new technology. On the one hand, our human body is not changing so much. The space defined by our human body is still there, but at the same time, if the whole world could have such a new intelligence in itself on the desks, on the buildings, on the whole city, then I don't know how we would define the relationship between our thinking and their thinking. It is something like the third existence, in a sense.

I'm just curious – I'm not positive, I'm not negative. I'm just curious how our perception, how our human body, could be adapted in a new environment. In such future situations, it's quite interesting because the digital has a sense of scalelessness, but our human body always has a sense of scale. Harmony of such huge diversities with different scales could be reinforced by digital technology; then our perception itself should be updated, and the whole urban situation and living environment could be also updated. Hopefully, we could have more richness of scale, in a sense.

When I am thinking and designing buildings, I am really a part of the building because in my mind, I'm in the space. But when it is finished, it is not me. The perception, interactions, and understanding of this one space could be so much more various.

I am curious about outer space, because I am always fascinated by how people think about how the whole universe started. I'm always fascinated by such challenges of unknown things, but as for physical experiences of outer space, I don't know – it's a bit scary for me to think about such an emptiness and extreme quietness.

Maybe, in a sense, I feel that space is densities. If I say forest, it's like the field of varieties of the densities, and the city of Tokyo is also something like that. Every time I design architecture, when you put a wall, then it is making the densities of the spaces where people feel safer or more comfortable or something. All architecture,

I could say, is like handling the densities of the air, in a way. The air gets really dense, then it becomes like the wall, and so on. In that sense, the space outside of the earth – outer space – is like the lack of density, and I'm scared of that.

ON SPACE:

Bilge Demirköz

Space right now to me means an endless sea of possibilities, of probabilities, of particles jumping in and out of existence. It's ever-expanding. We wonder what fills it. We wonder if we've just discovered the Higgs field, and it fills it? We wonder if there are any other fields that fill this vacuum, this space, whether it is going to be expanding forever, or at some point will stop and will deflate?

When I was a child, I would go on the swing by the Aegean side close to my grandmother's house by the sea. On the swing, every time I would swing up toward the stars at night, I would imagine that I was going toward space, that I was on a rocket going toward space. I think this was my first memory of the vastness, that I was this tiny little thing and I really wanted to be among them.

I'm absolutely fascinated by the concept of space – this whole encompassing concept in which we live. Where all objects move in, where all physics particles are in, and we're in. Is it empty? As physicists, we are questioning whether it is empty, because we now know that inside this thing that we cannot fathom – we understand now – there are fields of forces.

There is, for example, the Higgs field, which gives mass to a particle that might pass through it. We also understand that there is a soup of possibilities: at any point in space and time there could be a pair of extremely small particles that pop in and pop out of existence for an extremely short amount of time. This is really weird. It means that we have what looks like the surface of soup. Tiny little bubbles appearing and disappearing all the time which actually change the property of this soup. Does that mean that there is nothing? If you measure it directly, there is no real particle, but then there are these particles that we now call virtual particles sitting in space.

I'm very interested in particles and space. We have an experiment up on the international space station which is in space but only 350 kilometers away from us. It is the closest bit of space to earth actually. It goes around the earth many times per day and it sees many days and many nights in one day. It's up there being maintained by six astronauts, but our scientific payload built at CERN is called Alpha Magnetic Spectrometer (AMS) and it is like a sieve for particles in space.

There are a lot of extremely high-energy particles in space, and these particles are going very close to the speed of light. When they come to earth, our detector is like a sieve: it sees them, it selects them, it records them, and that sends the data back to us, for us to see it, for us to understand what these particles are. In some sense, this particle detector in space feels like part of my body. It's a sixth sense, so it's very interesting to understand these particles, because they are messengers from the rest of the universe. They are particles being accelerated in extremely high-energy events like supernovas and black-hole mergers. Then these particles come to us as messengers, having traveled millions and billions of years from wherever, deep in space.

There are three things about space that I will be extremely excited about if we discover them. The first one is the particle nature of gravity. We know about the dual nature of all the other forces in the universe [particle and wave], but we only just recently discovered the wave aspect of gravity, and we hope to now also find the particle nature of gravity.

The second one is dark matter.[1] Dark matter is the dominant form of matter in the universe, and I would really be excited if the Alpha Magnetic Spectrometer or the Large Hadron Collider discovered it. The third one that I'm really excited about is dark energy,[2] which I hope we understand, because it is the reason why we think the universe is expanding and we want to find out why. But maybe even the question is wrong.

We're solving some problems in physics, but, as I said, we're doing it only locally in the earth's vicinity. The question is: Is what we have found universal? For example, could the speed of light be different on the other edge of the universe? Often, we equate space with the universe, but then, who knows, maybe space extends beyond our universe. Maybe there are other spaces. Our understanding of space is constantly growing, and maybe it will continue to grow. That's scary.

There are lots of questions in my head. I think that space for a physicist is still an ever-changing concept. I don't think that we have nailed it down. It's still full of questions.

[1] Dark matter is a theoretical form of matter that is thought to make up 27 percent of the universe. The majority of dark matter is thought to possibly be composed of some as-yet undiscovered subatomic particles – like WIMPs (which is shorthand for "weakly interacting massive particles") or axions.

[2] Dark energy is an unknown theoretical form of energy which is hypothesized to permeate all of space, tending to accelerate the expansion of the universe. It is thought to potentially contribute to approximately 68.3 percent of the observable universe.

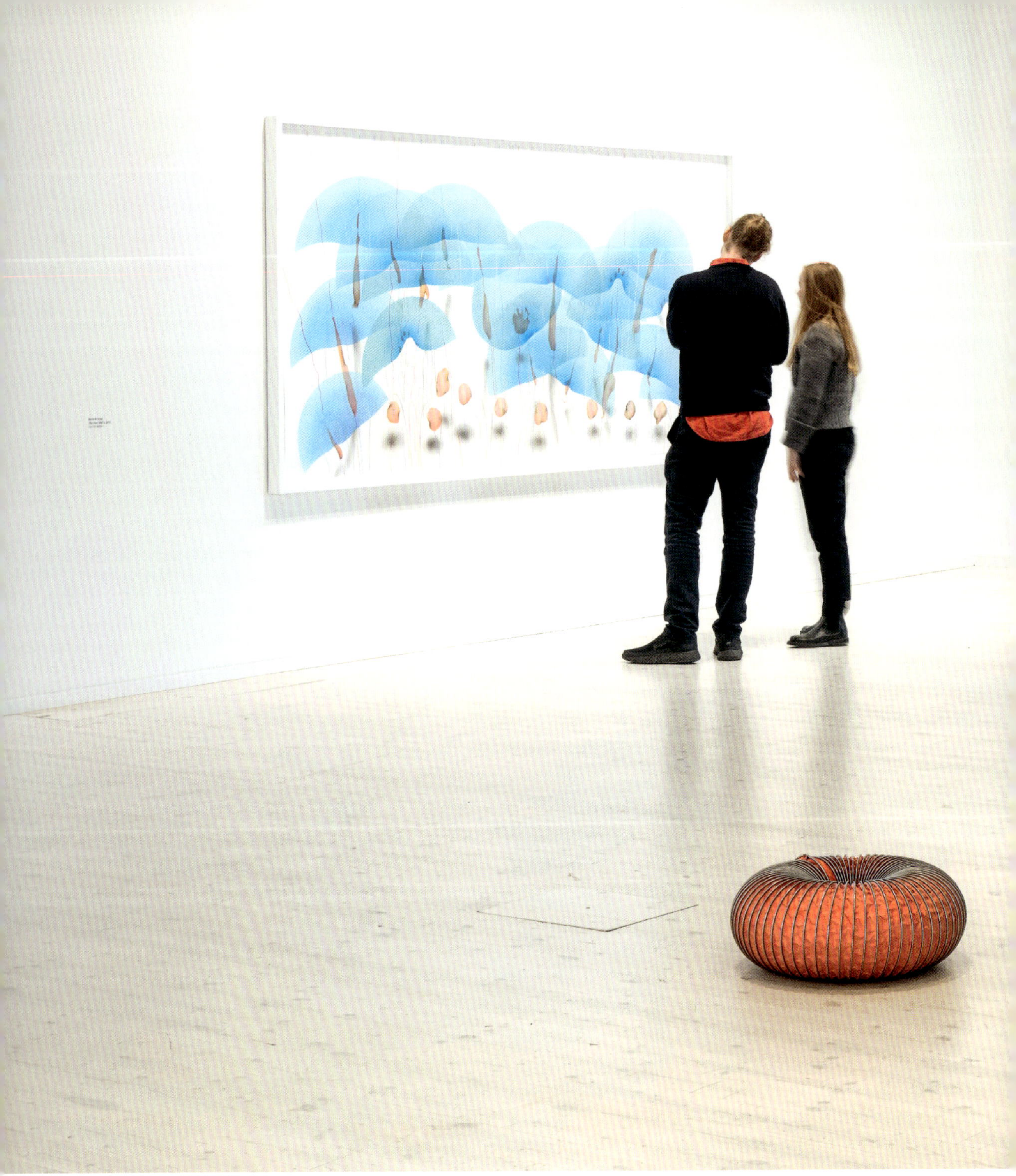

ON WALLS: *The Blue Shift I; Space Study 1–4*
Jorinde Voigt
2017; 2016

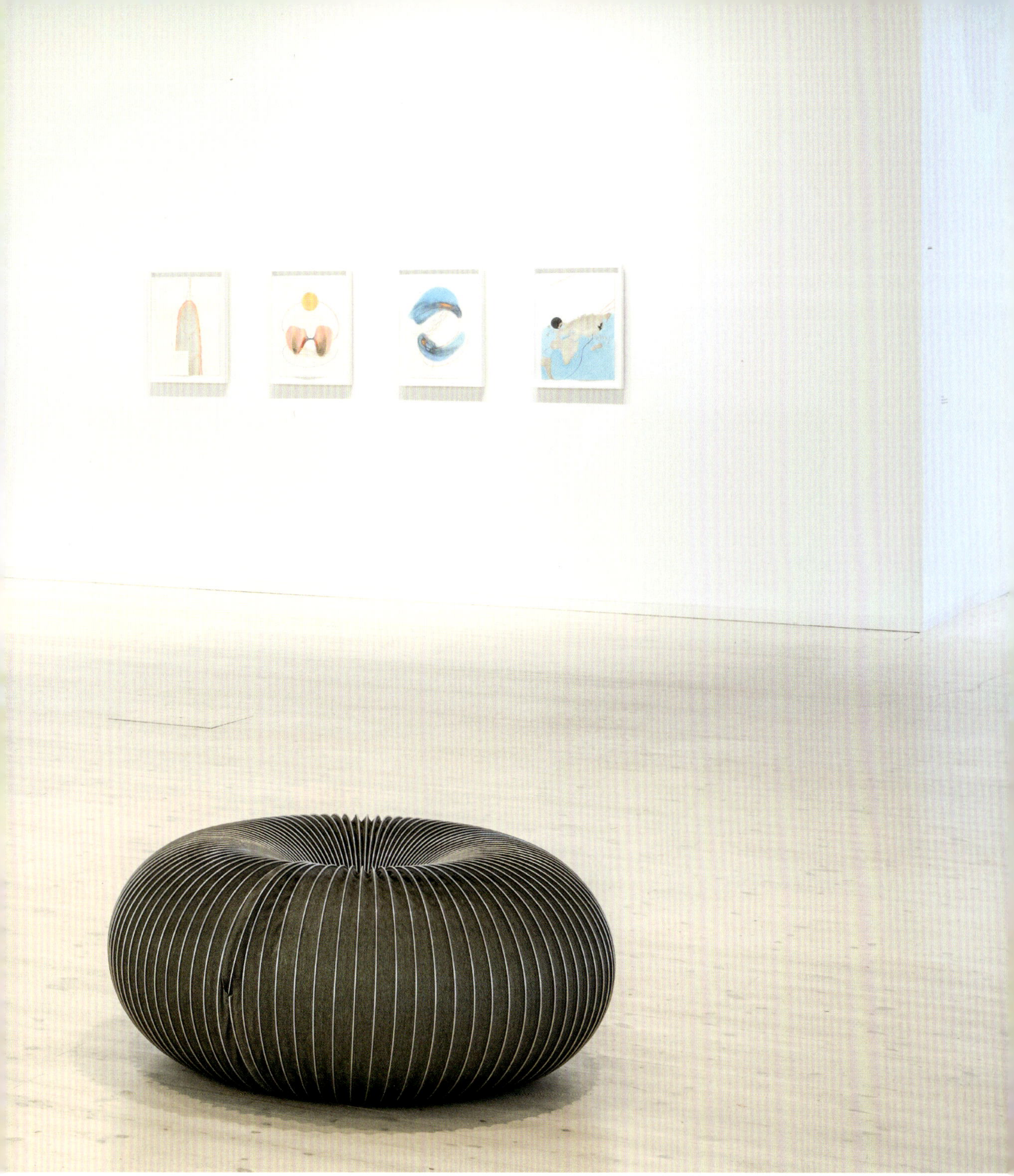

ON FLOOR: *Two Self-Revolving Toruses*
Julius von Bismarck
2010/2018

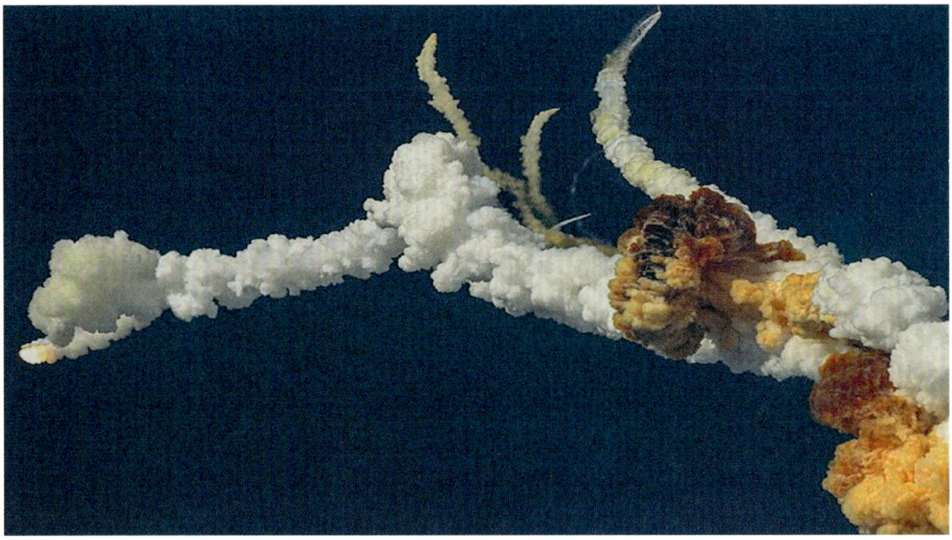

Still images from the film assemblage
Escape Shapes
Julius von Bismarck
2016

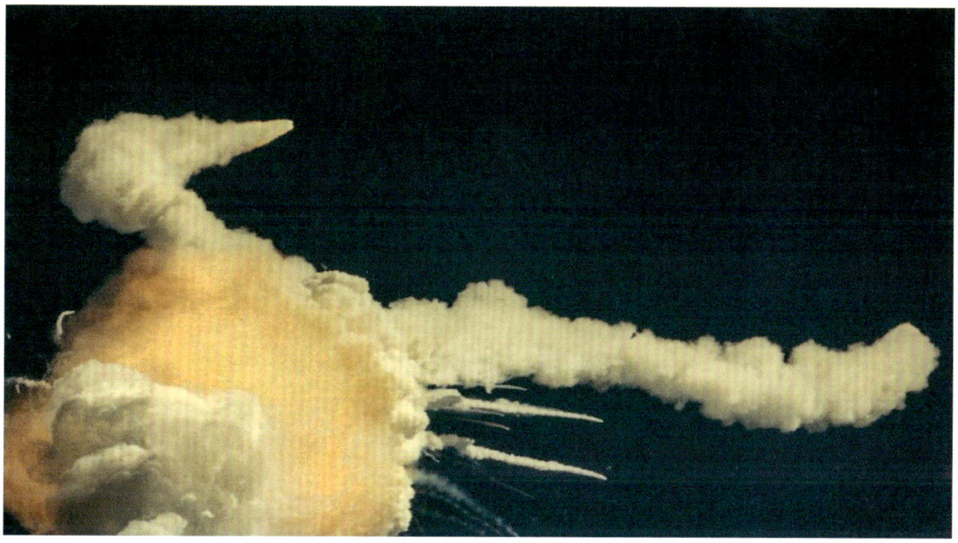

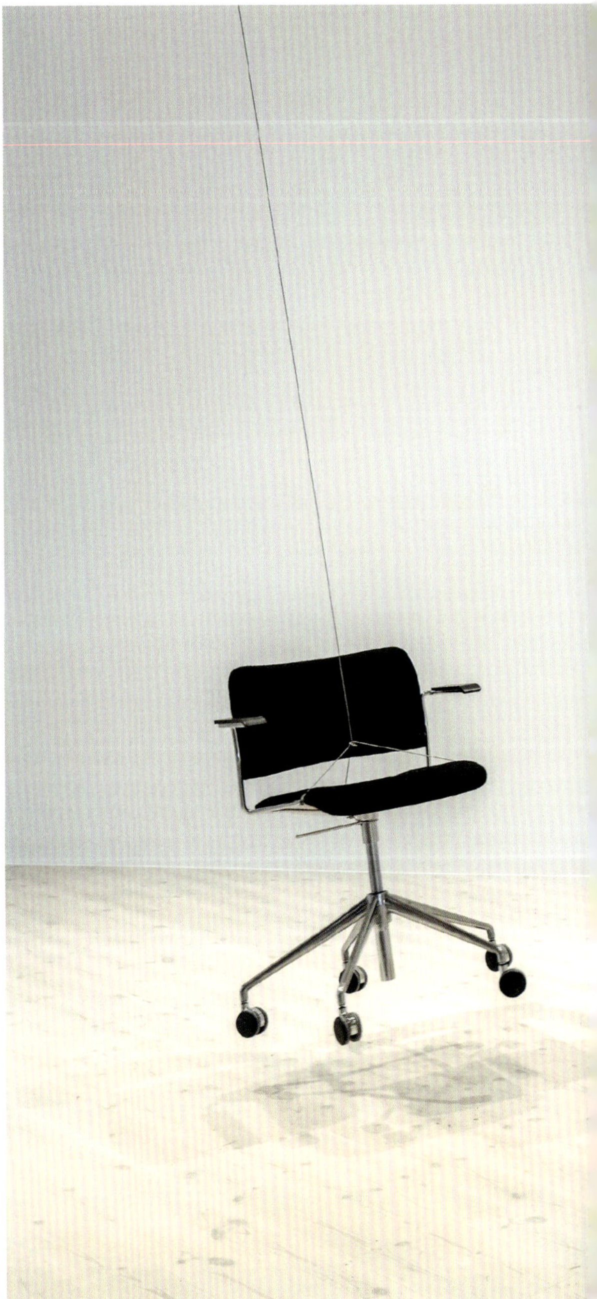

Freedom Table and Democracy Chair
Julius von Bismarck
2016

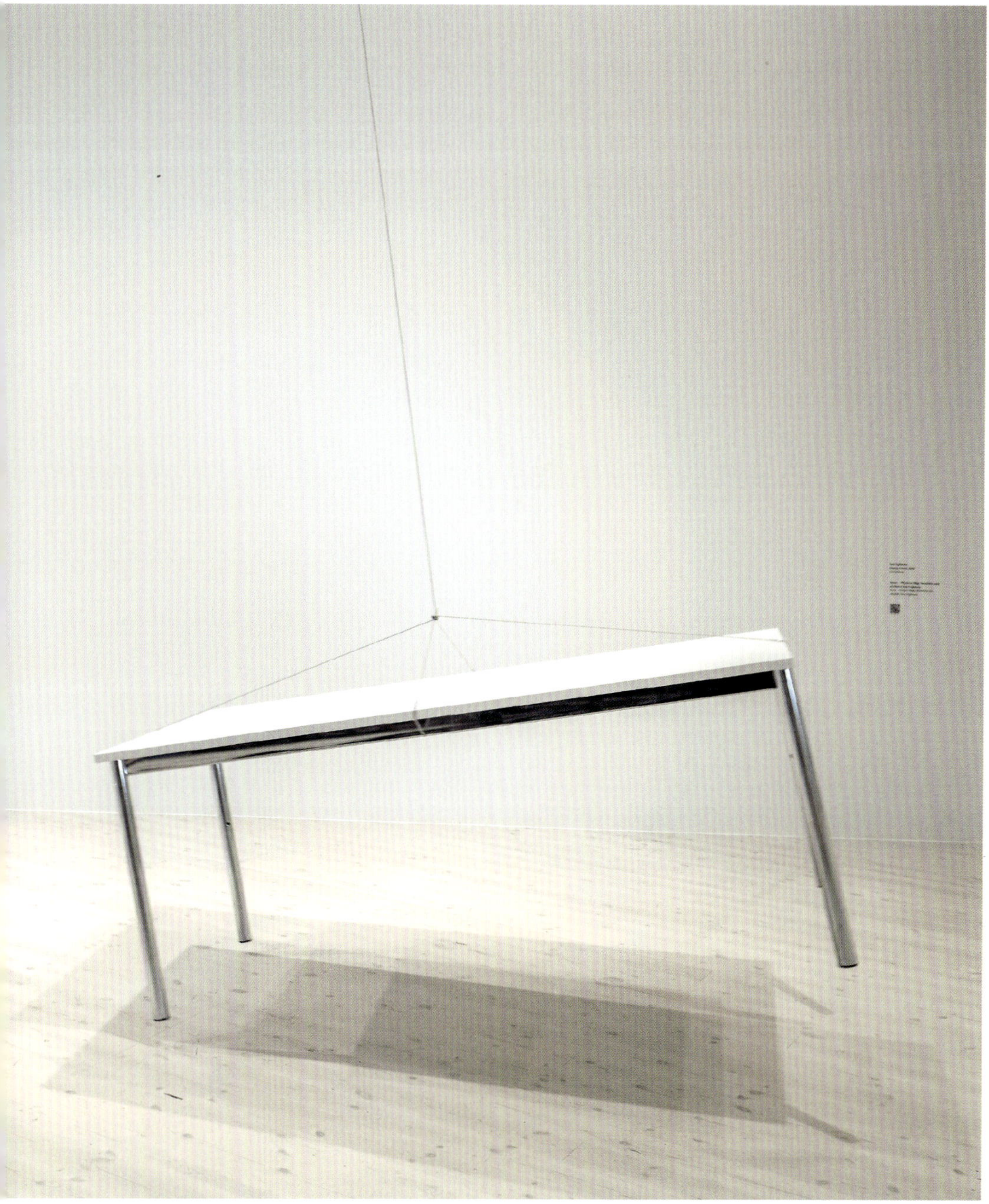

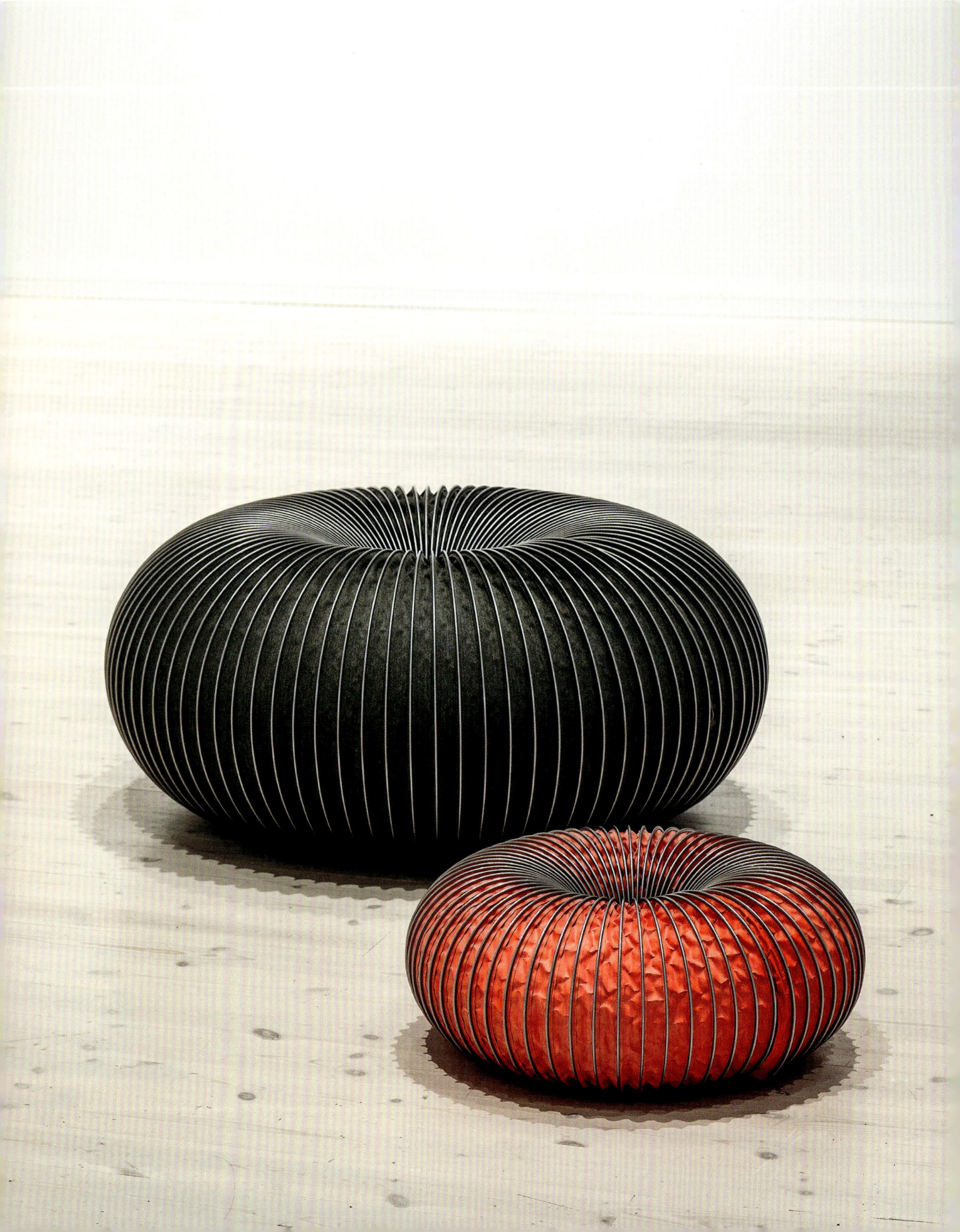

ON GRAVITY:
Julius von Bismarck

When I hear the word gravity, the first thing I imagine is a G actually. Maybe because of the g-force, but it's a nice warm thing. I also like the gravity pulling me down in my bed. It feels nice. It's a pleasure I like to feel. I think that I discovered physics without knowing the word physics. I was falling from a tree, that's the hard way of learning of the laws of physics. What actually made me more interested in physics is the point in school where it starts to get complicated and where what you can imagine in your brain no longer helps you to actually get what's going on.

My grandfather was a physicist, an important sun researcher. He died before I was born. I heard a lot about him. His name was Karl-Otto Kiepenheuer and he built a couple of observatories to watch the sun. Somehow, he also managed to put his apparatuses onto planes and rockets used during World War II, because he made everyone believe that his research was so important that he could even get military equipment to do his research. That's also sometimes inspiring for me.

As a kid, gravity is a thing that you always dream of being able to get rid of. I always had this dream that I could fly, but not fly in a way. It was aerodynamics. Just fly without any gravity, like really lose your weight completely and just fly.

Later, when I understood more about gravity, it became interesting because it's the thing that is different on every planet. It's surprising to see how we are actually able to adapt to a different environment, which we were never made for and where we never spent time in our genes, never had to adjust to that. We are still able to actually be in a space station. I would love to experience that, but I didn't have a chance to do that so far. I'm still stuck on the surface.

Instead, I have done experiments, for example, the egocentric system where I was spending a week in a turning parabolic bowl with a diameter of five meters [*Egocentric System*, Art Basel Unlimited, 2015]. Just enough space for me to sleep and

Two Self-Revolving Toruses
Julius von Bismarck
2010/2018

work and experience gravity in a different way, because you can't get rid of it on the surface. It was very interesting to experience step by step how my body is adjusting to that new force.

Gravity is something I'm obsessed with, because it's so hard to get rid of it. If you just want to fly, you can't – you're stuck on the ground. If you wanted to just have something that you see, you can't take it because it's not yours. Maybe for a kid, it's a similar frustration coming from these laws.

Then, later in school, you learn what force it takes or what acceleration it takes to make an object leave the gravitational field of earth and how fast you have to accelerate something to make it fly forever. That's already very interesting when you imagine that there's an end to this. It's not an endless force. It's everywhere. It's just very close to us, but the further away you get from earth, it weakens.

One of the big sources of inspiration for my work is physics.

Physics might sound very uninspiring for some people because it seems so logical and simple, but if you look closer at it, it's everything but that. What makes it so inspiring for me is that you can always discover something that you couldn't understand or even imagine before and that you maybe still can't imagine. This boundary of my imagination is where I get a lot of inspiration. Every new weird fact that is funny to imagine or hard to imagine is for me an inspiration to go into a corner of my brain that I haven't used before or that is not existing yet.

I have to kind of extend my imagination to understand it. Then, I'm failing in that. I like that failure. That's actually an inspiration for a couple of works I made. You can't challenge gravity. At the moment, you can get away from it – you can go into parabolic flight, zero gravity flight, where you are in a plane that flies a parabolic curve. Then it feels like you have no gravity. Also, I did this work where you have actually more gravity or it feels like you have more gravity. It's something you can't really get rid of, but there are tricks that make your body experience it in a different way. Scuba diving is a bit like that because you don't feel the gravity anymore, since your weight is the same weight as the water around you. Gravity affects the water around you as it affects you, so it's no effect in the end. You can just float around. Sometimes you forget where up and down is, which is a really weird experience, because normally it's always so clear where up is and where down is.

ON GRAVITY:

James Wells

When I was a very young boy, my father was an engineer, and he said, "All my bosses are physicists. I want you to be a physicist." It seemed like the greatest idea in the world to become the boss of my father, so I thought, "Maybe I'll become a physicist."

Gravity is the first force that humans recognize. It's what makes things drop out of their hand and crash on the floor, it's what keeps the moon in the sky, it's what keeps the earth traveling around the sun.

We experience it when we fall down ourselves, when things tumble down from a mountain. We think of it as a very obvious force, partly because we were taught basic things about it when we were young. As we grow older, we hear about other forces – the electromagnetic force, the weak force, and the nuclear force (aka the strong force) – that keep protons together and neutrons together and keep heavy elements like oxygen together. We think that that's much more complicated when we first hear about it, but gravity is the most complicated and least understood of them all, despite it being the most intuitive at a young age.

I always tell graduate students when they enter graduate school in physics that you begin your career thinking that gravity is very easy, and then you're in your career hitting your head against the wall saying, "I don't understand gravity." Gravity is the least known force because it's got a different structure to it than all the other forces that we have. Other forces have things that are minus and positive. Gravity is more or less all positive, and that creates a mystery that requires deeper understanding that we don't fully have. That's a shorthand way of saying some more subtle mathematical problems that we have not solved in gravity, but it's reflective of the issue of the difficulty of gravity.

A lot of people will try to solve gravity by thinking of very complex, deep, new ideas that we don't have intuitive feeling for. For example, string theory, where everything is made of vibrating strings like on a violin and every particle is one of those particular vibrations of the string. That's a simplification of the idea, but one works on it and develops a fast structure just to solve the problem of what is gravity. Nobody has found really a simpler way of approaching it. That's what makes it extremely difficult for us to get our heads around and to understand. I think that gravity and the difficulties of understanding gravity will stay with us for our entire lives.

One of the unique aspects of gravity is that it determines how many dimensions we live in. Electromagnetic forces might be just in one dimension or in two dimensions that could be felt. Gravity spreads out through every dimension, every spatial dimension that there is. That is the reason why in my research I was so interested in gravity, because it can go places that other forces may never be able to go.

One of the research projects that I did was to look for evidence of new dimensions by seeing gravitons being produced at the Large Hadron Collider at CERN. Now, the chances of that happening for certain were very, very low. We knew that when we did the theory work underlying that search, but it was a possibility that only gravity could afford and that is the reason why I became so interested in gravity.

Gravity also presents extremely interesting ideas for us to grapple with in black-hole physics. Black holes are very strange objects produced by extremely strong gravitational fields that seem to rip apart space-time itself. It's so strong and that is not something that we've ever known how to deal with. We are expending much theoretical effort trying to understand the properties of black holes and how black holes might affect other areas of physics, how it might affect the other forces.

In fact, recent experiments have found two black holes that were rotating around each other. These extremely massive objects, much more massive than our sun even, in a tiny space are rotating around each other more and more violently and vigorously until they merge and explode into a vast array of gravitational waves that travel through space and finally hit our detectors here on earth. We are able to see that because we thought about gravity so intensely for hundreds of years, and we were able to predict that such a thing could happen.

I think that the future will hold many things that I could never anticipate. The future in our understanding of gravity might have aspects that we never could anticipate. For example, right now people are trying to build a quantum computer.

A quantum computer has an entanglement of states inside of it that enables a parallel processing of very difficult problems that a regular computer could never do. You say, "Well, what does that have to do with gravity?" This is a computer on a desktop. You're using quantum mechanics from the 1930s, and now you're saying that gravitational research at the highest forefront level will have an impact on that?

Yes, it might have an impact on this for the reason that Stephen Hawking anticipated that there is information content to a black hole. That information content appears to be radiating away in a completely random fashion that loses information. This seems to violate our understanding of information processing, which is a very relevant topic for quantum computing. Black holes and understanding where information goes in a black hole might help us understand how to make a powerful quantum computer.

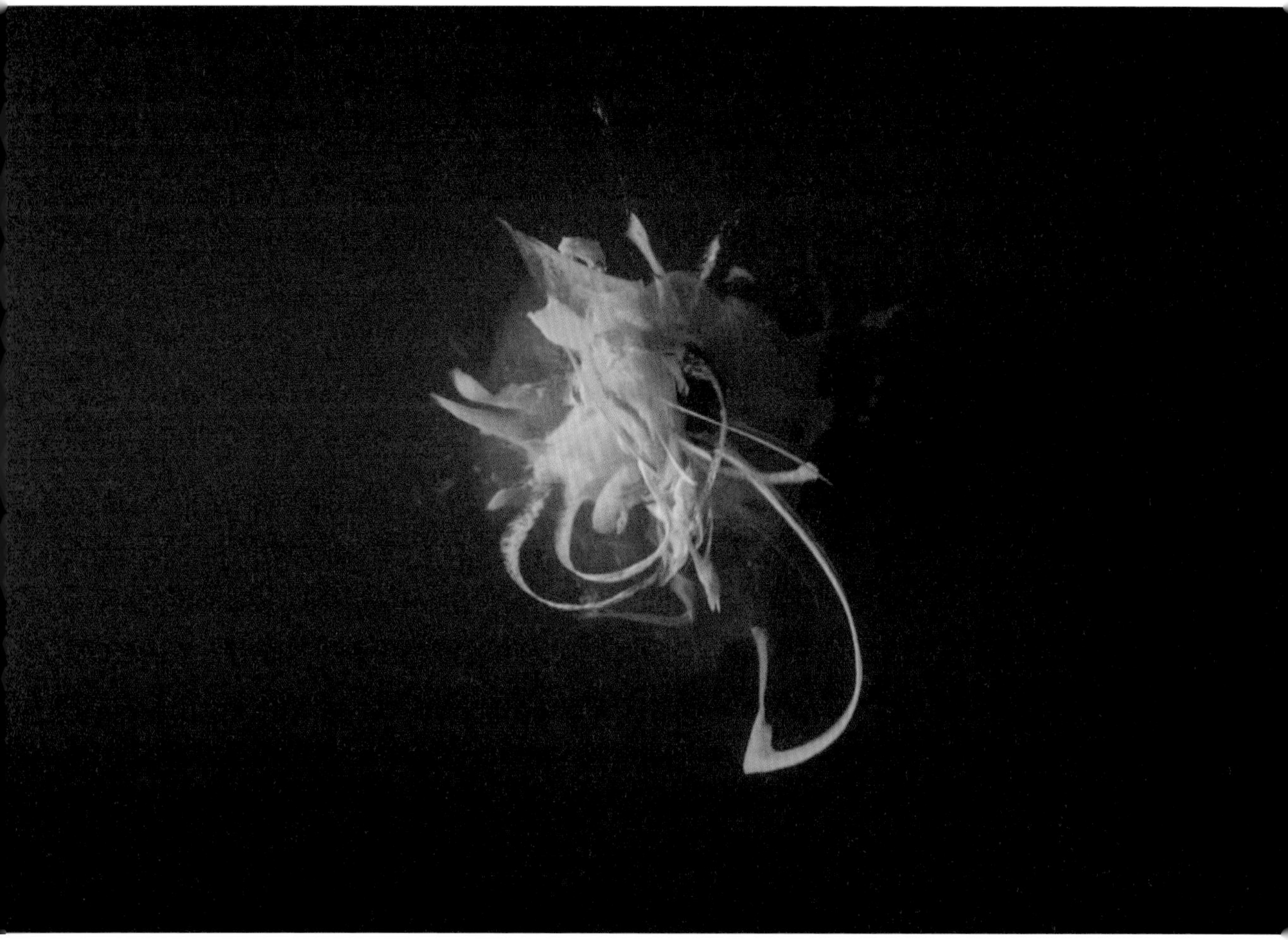

Still image from the video
Singularity
Solveig Settemsdal
2016

ON TIME:
Solveig Settemsdal

Archaeological dig by my house . . . open pit left behind . . . I'm walking out with a glass of wine in my robe . . . sitting in a Bronze Age pit . . . looking at the stars.

I try to make things that have duration rather than physical form. I am trying to compress time into objects effectively.

Time is a completely essential factor in all of my pieces, because I want things to exist in time rather than in space. For example, if I go to the National Gallery and I'm looking at a painting from 1600, then I'm imagining the pigments changing, compressing the time it took between when it was made and now.

I see time as very, very fluid, and very viscous. I was reading *The Inflationary Universe* by Alan Guth a while ago. He was describing the universe as initially expanding, inflating, and then cooling or coagulating like gelatin. I had this very playful idea of space expanding and kids' desserts and pooling liquid time basically. That really exacerbated my obsession with this material [gelatin, which the artist used to make *Singularity*], and with this attempt at expanding objects in space and time at the same time.

My work called *Singularity* was very much inspired by this, and also by thinking about something expanding from a single point, and then disappearing again into the same point. Between those two points, there's this life, this evolution, this growth, this striving to be and getting to a point, and then reversing. I think that my favorite point is where "it" goes from being perceived as something that is generated organically, something that happens – a drop in water – to where "it" seems to have its own volition and being. "It" seems to have a will, somehow . . . you're drawing in five dimensions effectively.

I was looking at Paul Klee quite a lot, and at "Taking the line for a walk," thinking about the flatness of paper and what happens when you start something . . . the first point on a piece of paper is like a thought appearing. Instead of using a pen, I'm using a syringe [filled with gelatin] which draws in the space. I can make a form that is constantly evolving, constantly expanding, but I am not drawing as in describing something. Rather, it is describing itself, and I tend to talk about "it" like this, like "it" is a thing of its own that I didn't make.

The origins of things have a massive influence – how things become things, what they go through, and that goes from putting paint to paper to how the universe began, and it just kind of expands outward like that, really.

I think that all of this goes back to this dream I had when I was really small. Every time I had a fever, I had the same dream, and it was really kind of elemental: a white sphere hovering in a dark space that would slowly morph into something really grotesque and then back into this pure beautiful thing . . .

That got me really thinking about why we make solid things. Traditionally, art is supposed to solidify and art is supposed to be eternal, and ideal platonic solids, and invincible monuments, et cetera, or things that make us feel like we're solids, that we have the civilization which has a certain structure, with certain values that we share. The problem is that we're not set like that at all. We're completely shifting, changeable, movable, fluid beings, and our thoughts morph constantly, but we have this hankering to make eternal monuments . . . and it is definitely changing with the Internet . . .

. . . I think about it [time] very much as a fabric, as tissue of a tissue, as material that you can kind of stretch and bend, and pull, and push. There is a way of beating time in the sense that if you can control the environment that you are portraying time in, then in a sense you can. But if the material just takes it, then it's just a boring game. You have to have a bit of a battle. If you can't play with it, then it's just dead and therefore boring. The forms are very much made by what the material can do, and also what happens. When something happens, then you respond to what happens rather than enforcing something onto a dead material that has no backbone.

I'm constantly surprised by things. I think that the element of magic is really important. When something happens and you're genuinely like, "I can't believe that happened at this time and I can't believe you can do that," that's where I want to be.

ON TIME:

Peter Galison

Time is something that for us evokes ideas of mortality and our humanity at its most essential form, but that also stands absolutely central to the ideas of Newton and Einstein and more recent ideas about the nature of the physical universe. It's that cross-connection that fascinates me, that doubling up between our human, very visceral, immediate understanding of our lives and our place in the world and the larger structure of the universe.

As a small child, what really struck me was that my grandfather, whom I adored, had a collection of clocks – fantastic and beautiful clocks mostly from late-nineteenth-century Paris, but also from Vienna, Berlin, and other places. Every morning he or my grandmother would walk around the house and wind them as if they were watering plants. I remember looking at them, at the mercury, pendula, and some of the clocks, or hearing the way they clicked or chimed or opened. I found that absolutely fascinating.

I learned relativity before I learned proper Newtonian mechanics. I started out by thinking that it was normal for every observer to have his or her own time and own measures of space. [Einstein's Theory of Special Relativity, 1905, which is in opposition to Newton's idea that time is universal and constant.]

I thought that was completely the way things were done. In a sense, I've tried to compensate for that with my own students by really trying to have them understand what a shock it was to think that time was relative, that each observer in motion carried his or her own proper time. It's really amazing.

That doesn't upset people quite so much as the idea that time changes, because time is so strongly associated with our existence, the beats of our heart. The heart and the fact that a second is approximately the beat of a heart is very close to our life itself, and the amount of time that we're allotted on the earth is deeply ingrained in us as something that is associated with what it is to be human. If you mess with this, if you say, actually, that if your twin went out to a faraway star at a fraction of the speed of light and then came back, then your twin could end up younger than you – that seems amazing, disturbing, unnerving even. It's also [scientifically] true.

We test that all the time at CERN or in any other accelerator in the world. We see particles that normally live for a tiny fraction of a second living much longer. We see cosmic rays as particles that come from deep in space that by all accounts should be decaying into other things, but they reach the surface of the earth. The reason is that their clocks run slow, so to speak, and so they live long enough to reach down to sea level.

We see this in a million ways ourselves. We see it with our telephones and watches when we use GPS. If relativity wasn't right, then our maps wouldn't work. We'd get lost in our cars. Our watches would no longer synchronize. We test relativity all the time in our daily life in this way and in others. We test it through the motion of Mercury around the sun, but all of these are relatively weak bendings of space and time. Around a black hole near the event horizon itself, time and space are curved; they're distorted in a much more extreme way. That is what [an experiment called] the Event Horizon Telescope will ultimately probe.

If you fly from Europe to the United States and back, then you are younger by some tiny amount than somebody who stays at home.

We know that you can travel into the future, and if you imagine, as you easily can, getting rockets to go faster and faster, then you could end up moving into the future and living beyond the time when everything and everyone you know has long ceased to exist. Whether that's a dream or a nightmare is another question, but the idea of traveling into the future by no means contradicts physics. It's completely in agreement with the physics that we've known about since Einstein sent in his paper in June 1905.

We have well over a century of thought about the way time can bend in this way; it's also the case that when you stand on the earth, that your head and your feet age at different rates because being in a gravitational field distorts time and the progress of time. You're time traveling even one part of your body relative to that of the other when you stand up or stand on your head. That is well accepted. The idea of traveling backward in time is much more disturbing, and it is much less clear that it's compatible with any kind of physics.

Look at the history of art: if you go back and you look at a Renaissance painting and you see a sundial or a sand clock, then it's always talking about mortality. It's always about life and death, the possibility of an afterlife, about these fundamental aspects of the human condition . . .

Clocks are never just clocks.

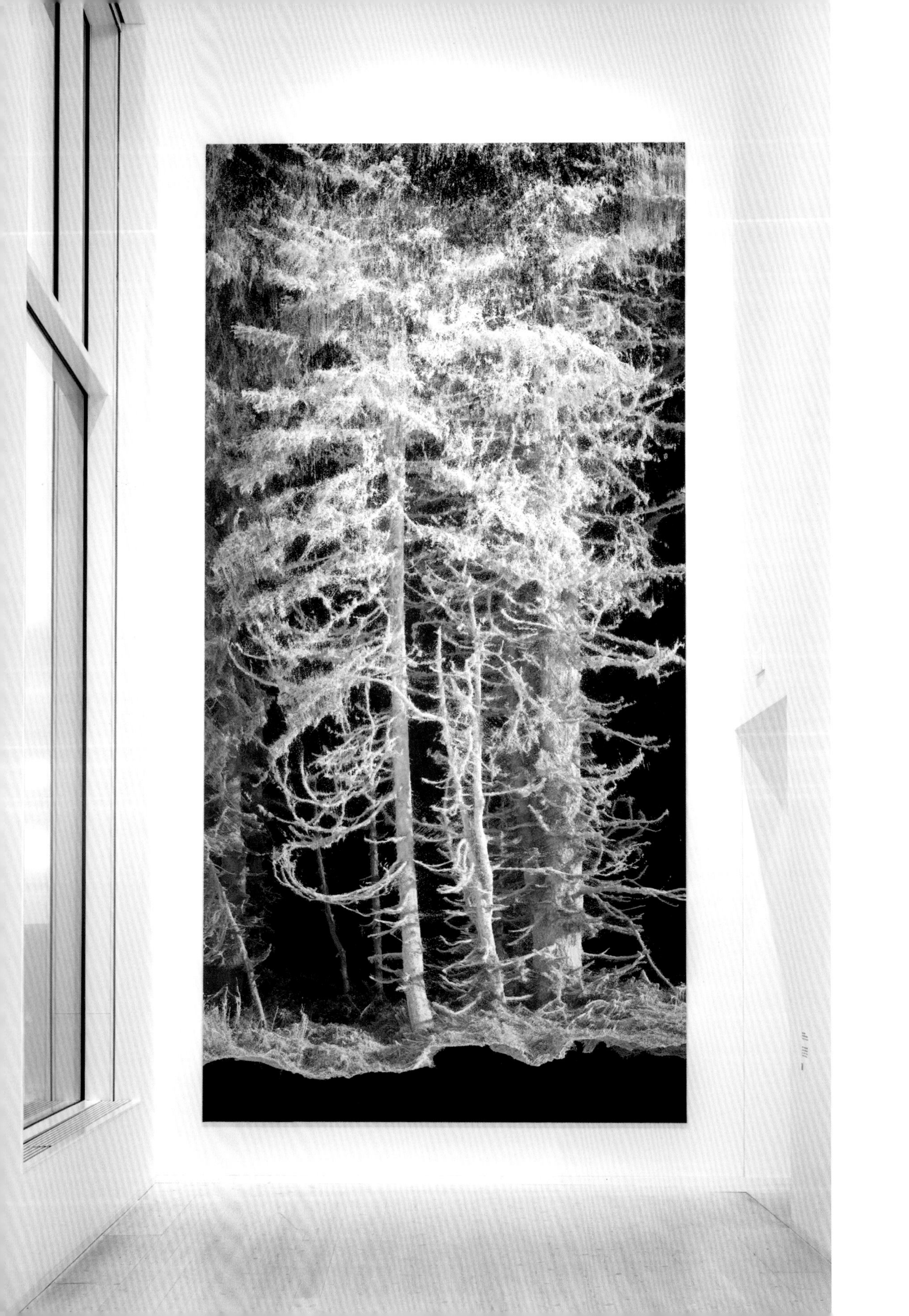

ON LIGHT:

Davide Quayola

It is strange what kind of role light plays in my work. I think of the contrast between the background I have had growing up in Rome, dealing with a classical visual culture that is very much physical, and what I do today – it is very ephemeral. I deal with light almost as a material. It's something that I can capture, can measure, and can sculpt with ultimately.

For me, light is a material like paint. It is something that's in my work; it is captured through different types of sensors, different types of machines perhaps, but also it is something that is measured, that is sculpted both digitally but also somehow physically when I do my installations.

This brings me to an important aspect of my work, which is the relationship between physical and digital or between material and immaterial. What I find fascinating about light is that it is both immaterial and material at the same time. That, I think, is particularly what I'm trying to do with my video installations – to create objects and works that are both physical and not physical at the same time.

I'm very interested also in light as a source material that can be analyzed, that can be measured in a very different way from how we perceive light. Technology allows this different set of eyes that expands our senses and allows us to see light or somehow understand light in different ways under different conditions. For example, my work *Remains* is a large-scale print. It is indeed a physical object, but it is the result of a rather complex computational process which is a very sophisticated measuring system. Essentially, it's trying to capture distance between this measurement device and the natural features around it. By capturing these distances, it is able to re-create 3D-accurate visualizations of the surrounding nature.

Again, it is all based on light, on measuring light, on measuring distances between these objects, but again it results in something very, very different than a photograph and than our sight.

Remains
Davide Quayola
2018

The name *Remains* really came while I was exploring these spaces – not physically while I was capturing the data, but digitally when I was exploring the data. Then I started to realize that what was really interesting was where the parts were half fully captured and half perhaps broken due to some kind of errors, also maybe because of the limitation of some of these devices to capture something so complex in its geometry. I started to focus on parts that were both completed and quite realistic, but at the same time also broken and noisy, that almost became painterly. I think that this is why there is this aesthetic that is in between a photo and a painting, because it has these kinds of errors. The impossibility of the machine to capture these features perfectly becomes almost like a paintbrush of the artist that develops its own aesthetics.

I would say that the idea of working with light today for me is not necessarily a choice. I think that it is becoming a material for us as a society: it's how we deal with screens and how we deal with the digital and with data in general. I think that our relationship with light certainly has extended greatly, over the last thirty, forty years. For me it wasn't necessarily a clear choice of being an artist who works with light, but just a natural reflection of living in this time.

There are several dreams that are stacking up which I want to do in the future. One of which, to me, means really bringing down this boundary that divides the painting from the video.

In the future, I believe that there will be possibilities to both capture and output light as a volume, and to work volumetrically rather than on a surface. I think that would create incredible opportunities.

Without light, there's nothing, there's no color. You cannot quite separate light from any artistic process – almost. Light is clay. Light is data. Light is material and immaterial. Light is the input, light is the source. Light is the subject, but also light is the output. Light is THE material.

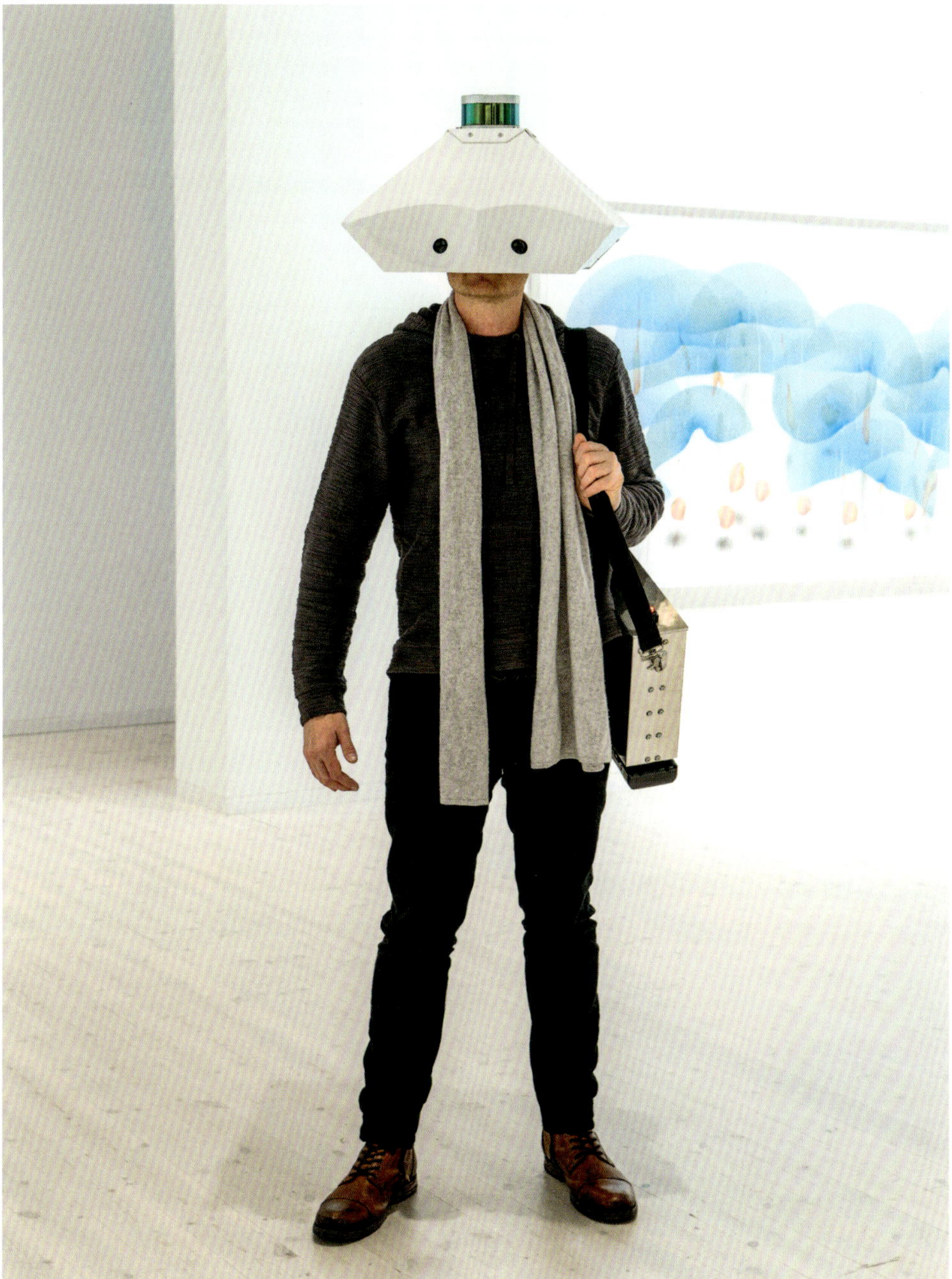

Lidar Helmet
Julius von Bismarck
2018

ON LIGHT:

Kate Shaw

As a physicist, light is a relatively well understood phenomenon. You emit light by an oscillating electron. When you oscillate an electron, it has an electric charge and it makes electromagnetic waves.

When we talk about light in physics, we actually mean the full electromagnetic spectrum. Whereas in everyday life, we might just really talk about light as the narrow width of the electromagnetic spectrum that we can visibly see and interact with. Of course, there are also other wavelengths of light that we don't see and we don't interact with.

For example, when it's a dark night and you're outside, you might think that there's no light, but from a physicist's point of view, there's light everywhere. It's all electromagnetic radiation: it has different frequencies and different radio waves. There's radio waves, there's gamma rays, there's microwaves, there's infrared. You can also use different types of telescopes and types of ways of detecting the different types of waves other than just the eyes.

Light is very important for the universe. It's one of the key components. Light is also the quantum excitation of electromagnetic waves which we call a photon. In fact, almost everything that we see around us is bound together by molecules which interact and are held together by the electromagnetic interaction, which is actually light or photons being interacted between the atoms. Not only is light everything that makes up the universe, but it's how we experience, as humans, the universe. It's how we see everything around us today. It's also how we experience galaxies. It's how we understand that the universe is expanding, that it is accelerating.

When we talk about the formation of the universe, we always need to go back to the Big Bang. The Big Bang occurred around 13.8 billion years ago, a long time now. When this happened, all the energy and matter came into being. One of the key parts of this energy were the photons, which is light or electromagnetism.

As the evolution of the universe occurred, there was a point at which the photons could then travel across the universe without having to interact directly with particles, which means that nowadays we can actually listen and hear some of these photons that came from the beginning of the universe. This is called the cosmic background radiation. In the formation of the universe, light or the electromagnetic spectrum is really one of the key four types of forces that we have.

I should mention that light is massless. It has no mass like you and I – we are made of atoms, electrons, protons, neutrons, and they have a mass. Light, photons of light within electromagnetic waves, don't have a mass. Therefore, they travel at the speed of light. Anything that travels near to or at the speed of light experiences time and space differently than us.

If I would somehow manage to reduce my mass to such a small point and travel up to the speed of light, then I would be experiencing time and space very differently than you. If I were holding some very tiny clock that had very little mass, then my clock would be going slower than your clock if you were rested on the earth and I was going at the speed of light. Time is actually a completely relative phenomena, and it's completely defined by this universal speed limit, the speed of light. Einstein's theory of special relativity told us that there is a universal speed limit which is 3 x (10^6) meters per second. That's the speed of light. Nothing can go faster than that speed. Physicists want to know if you can break that theory.

It's really fascinating how we perceive the world around us. When we look out at the night sky and we see the billions of stars shining down on us, we're not seeing what the universe looks like today, and we're not even seeing the universe as it looked last week – we're seeing the universe hundreds, thousands, even millions of years ago. The nearer the stars are, the shorter that time difference.

Day to day, when we look at each other, we are always seeing each other just a few seconds in the past. In general, it means that when we're looking at everything around us, we're just simply interacting with light that's hitting our eye now. We're not seeing everything that is around us. We're not looking at a tree a hundred yards away. What we're looking at is a photon of light hitting our eye that left that tree a certain amount of time ago. We're always actually only interacting with photons upon our eye. Everything that we believe we see around us is simply a reconstruction inside of our heads.

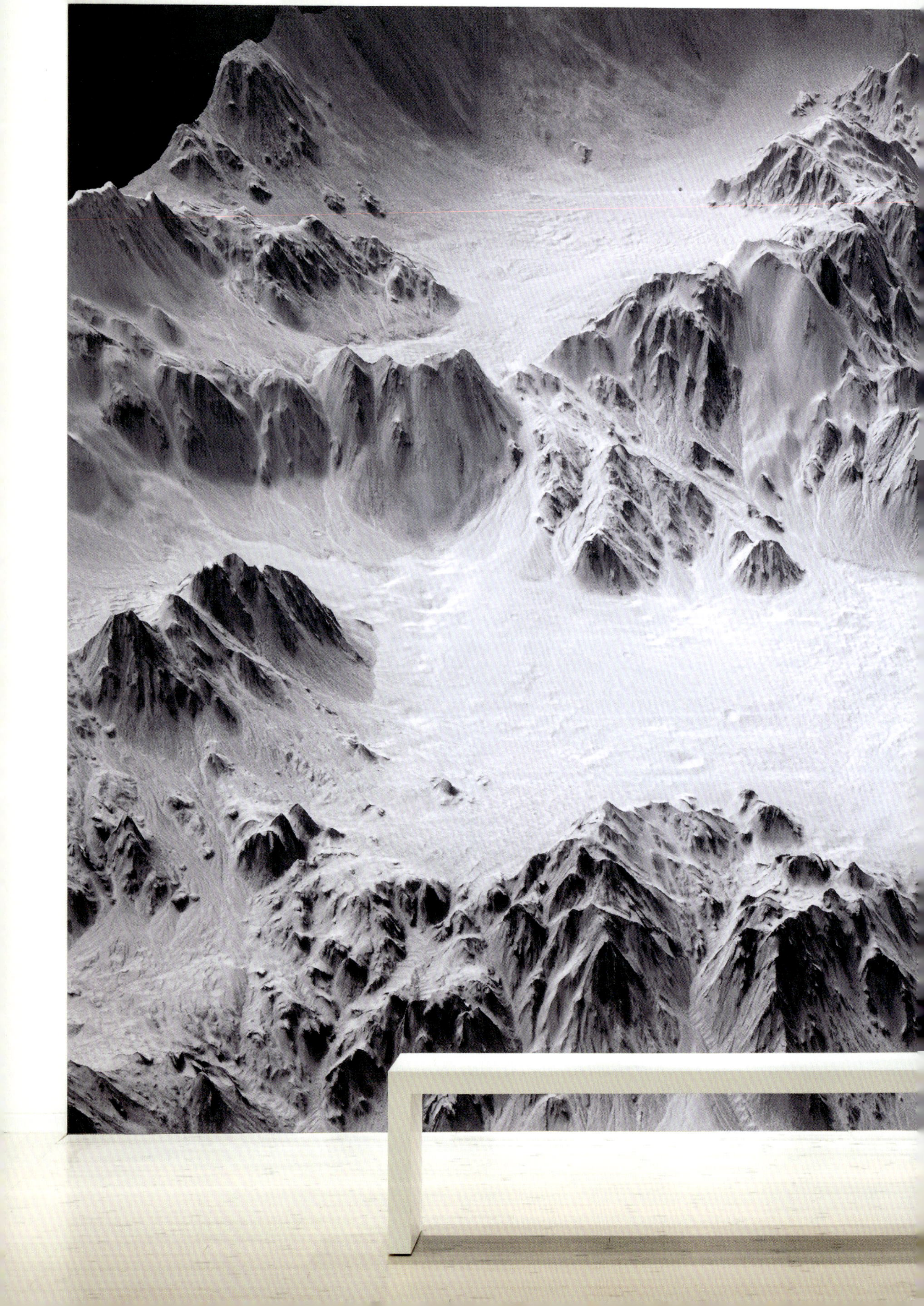

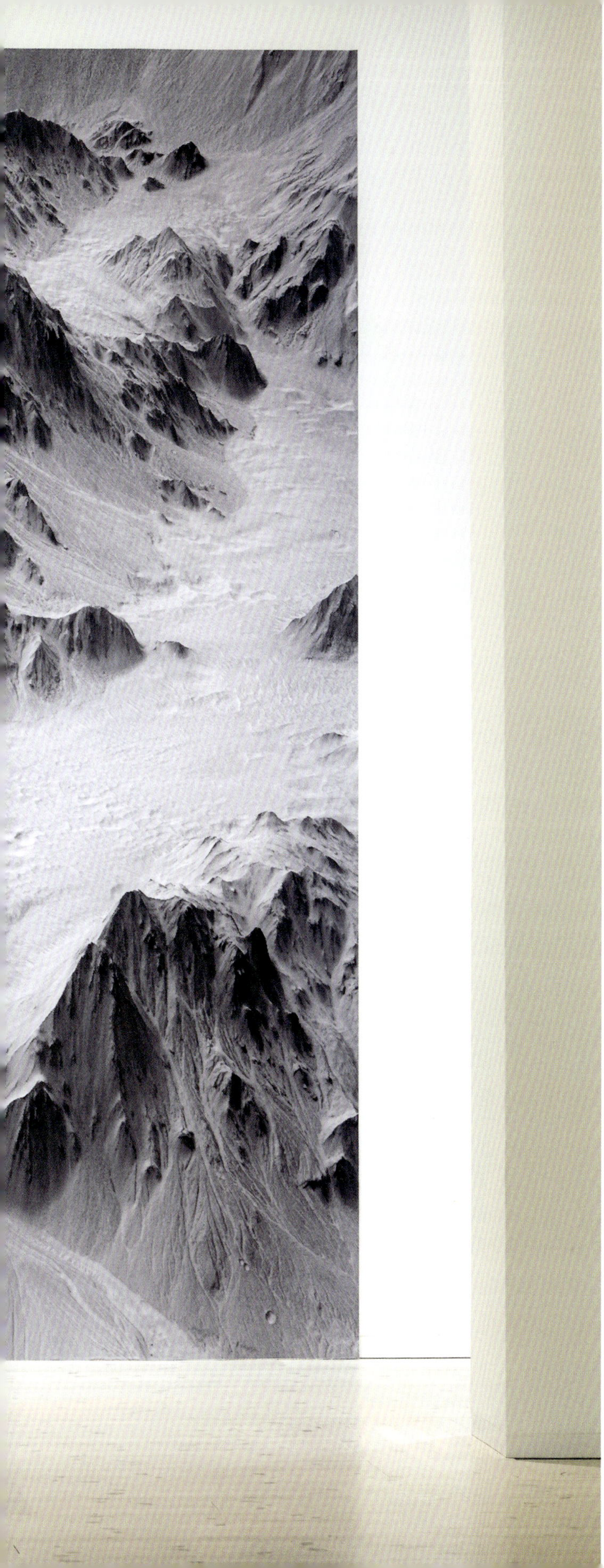

Contracting Universe
Carey Young
2010

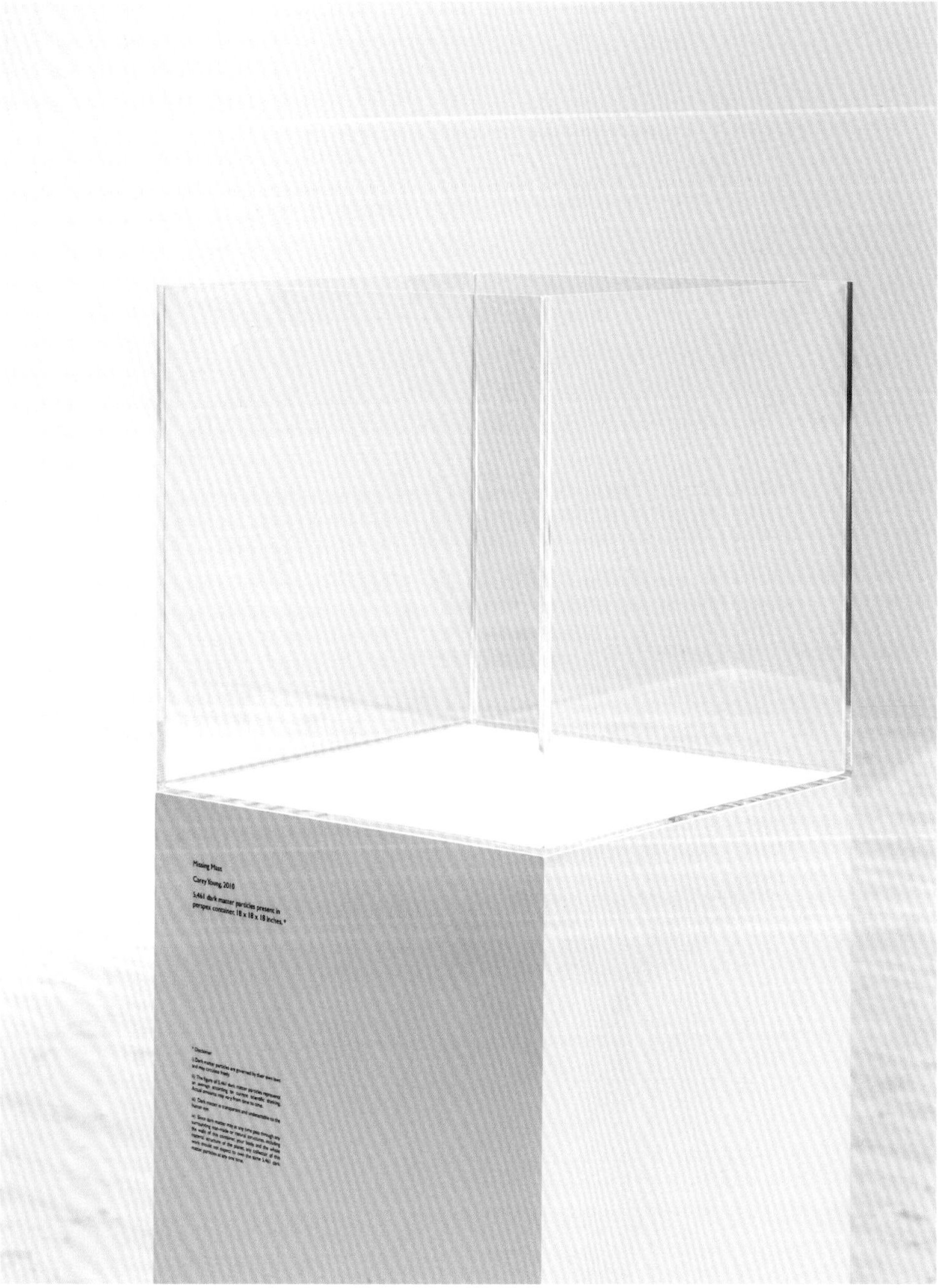

Missing Mass
Carey Young
2010

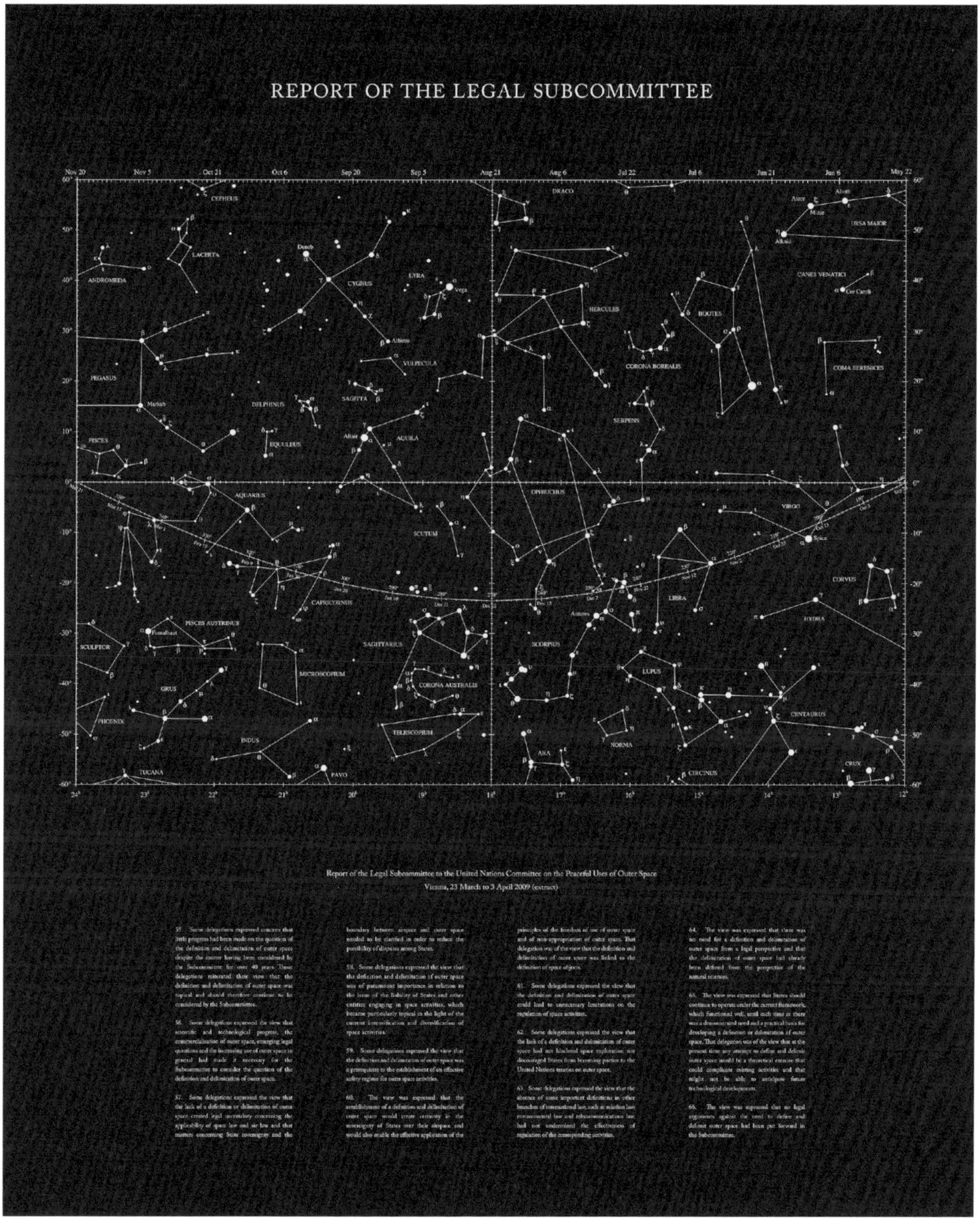

Report of the Legal Subcommittee
Carey Young
2010

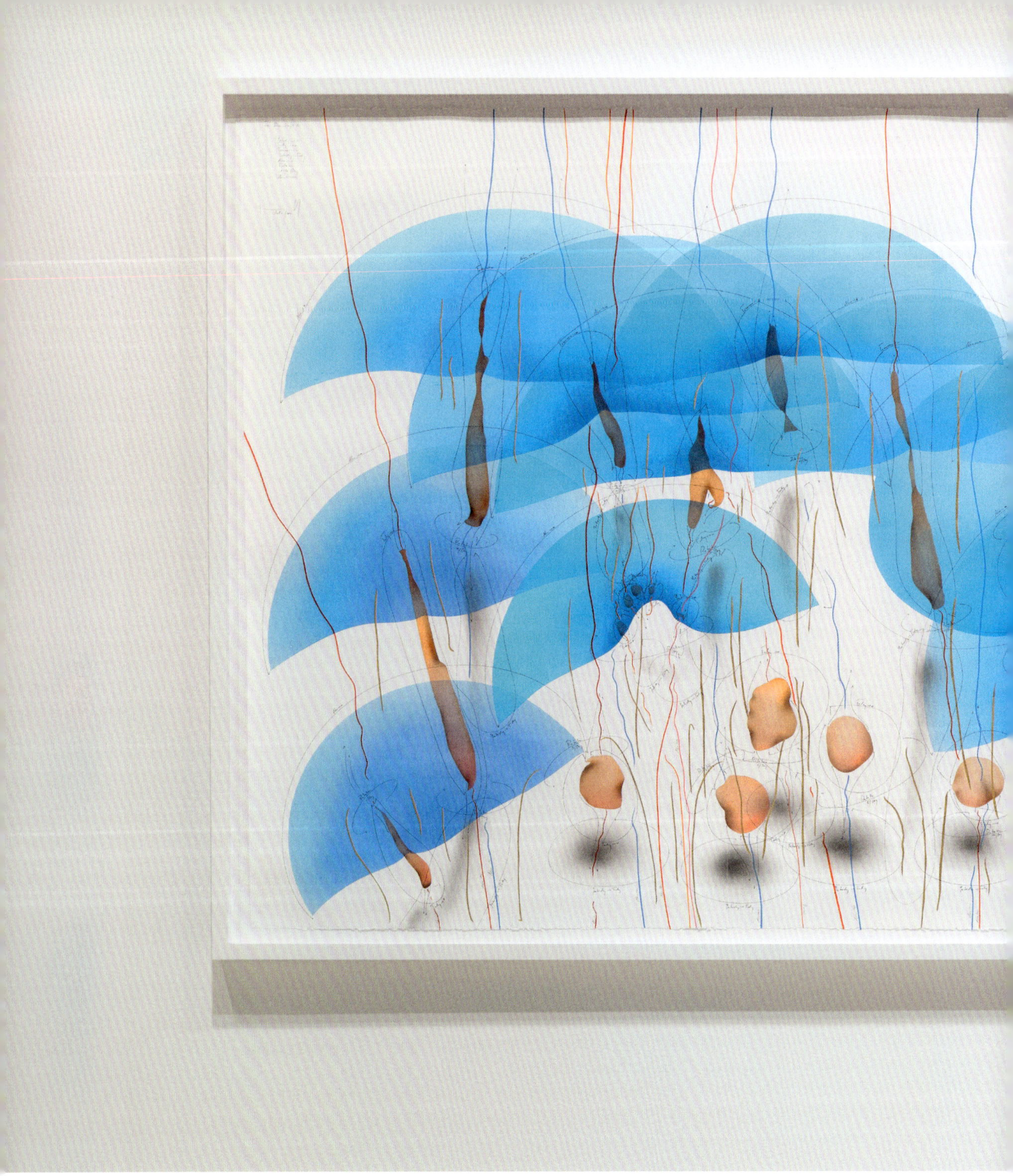

The Blue Shift I
Jorinde Voigt
2017

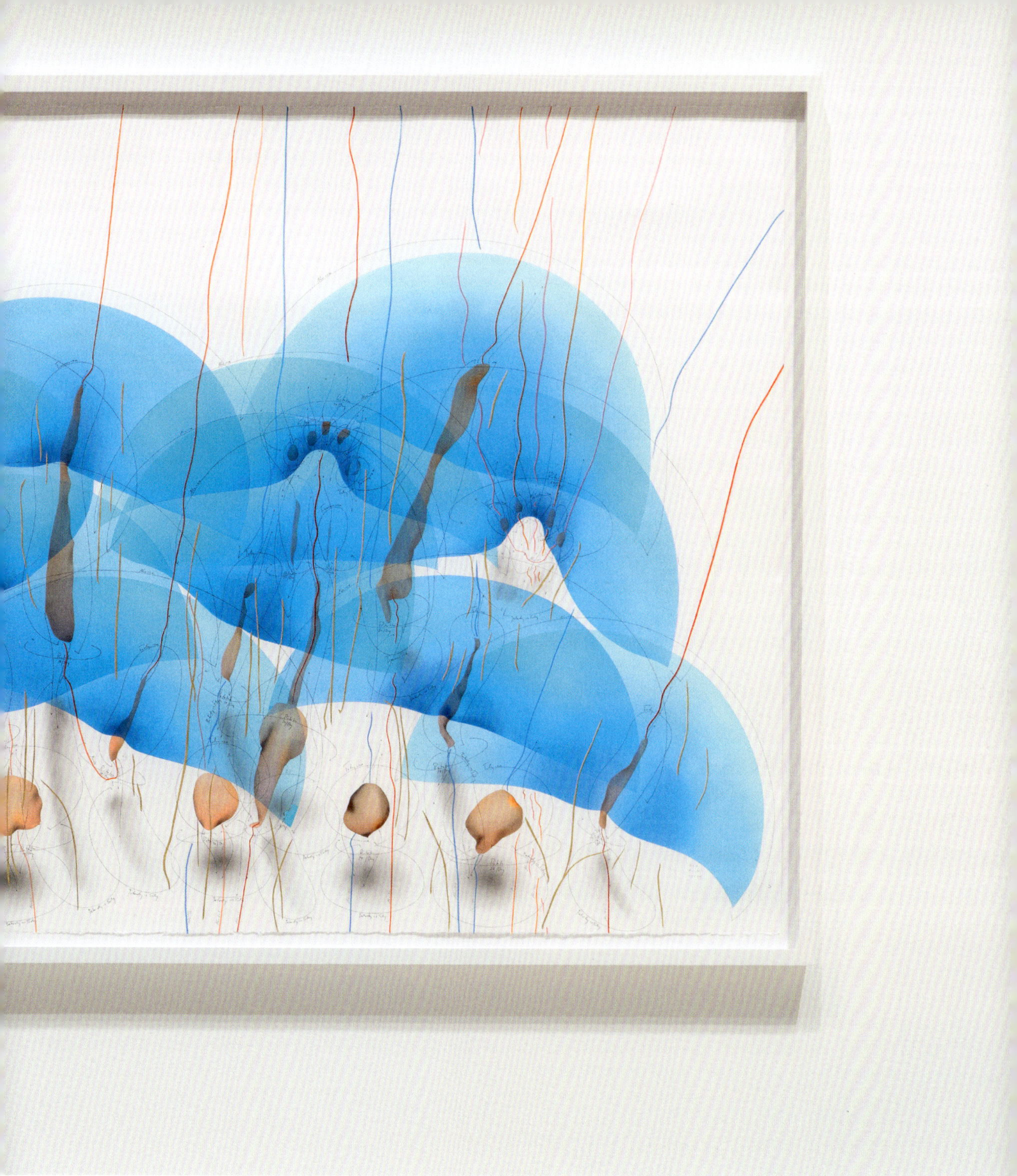

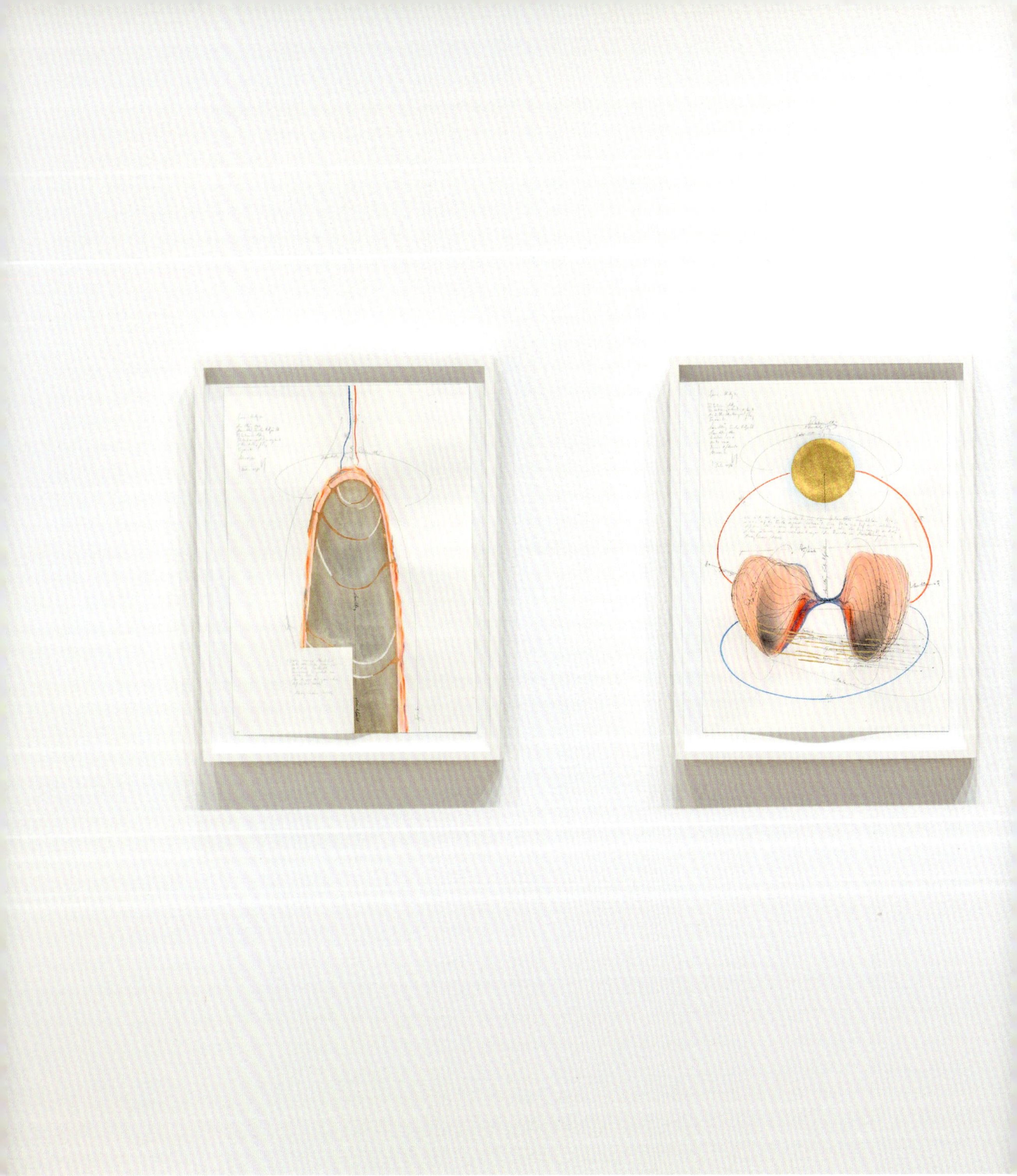

Space Study 1–4
Jorinde Voigt
2016

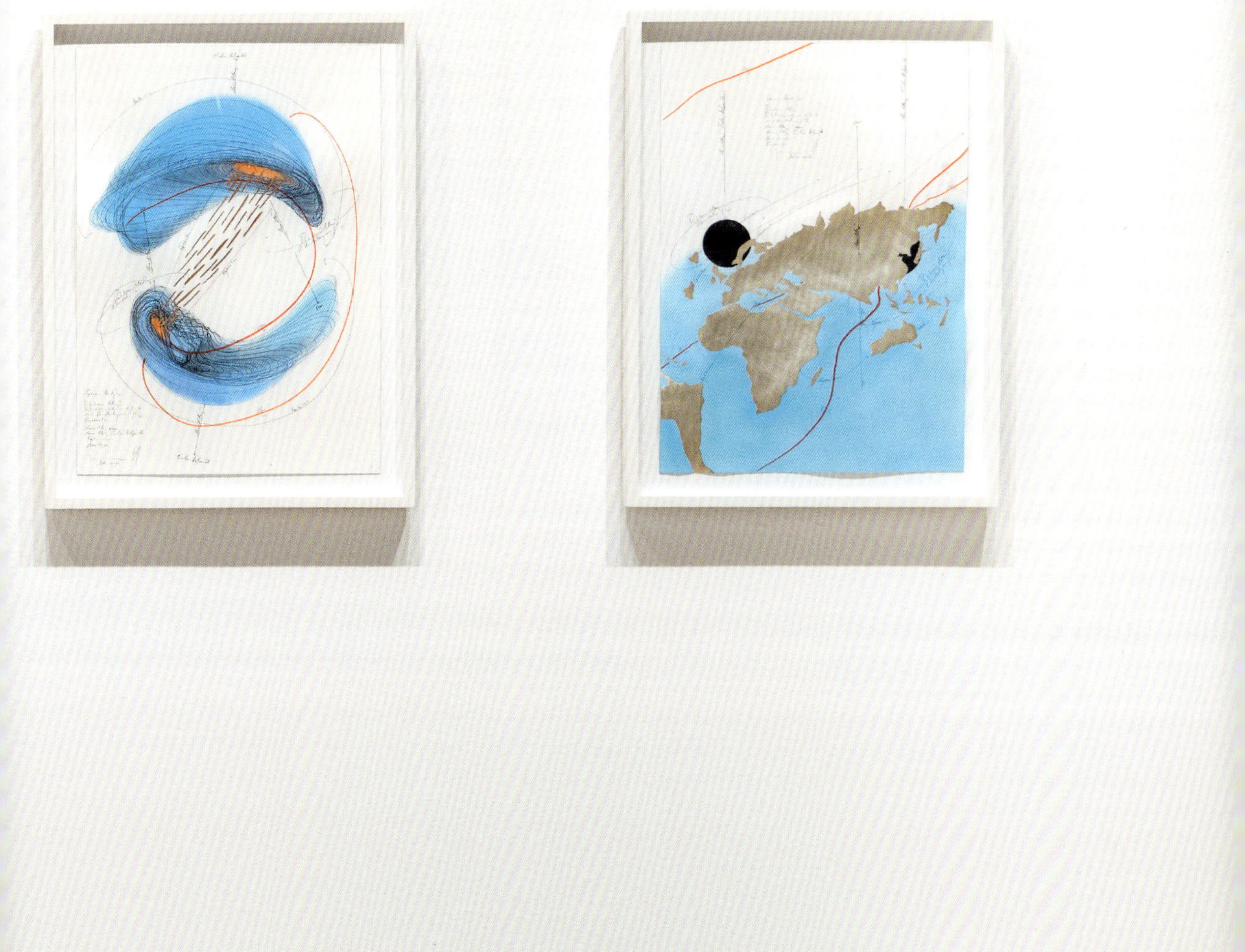

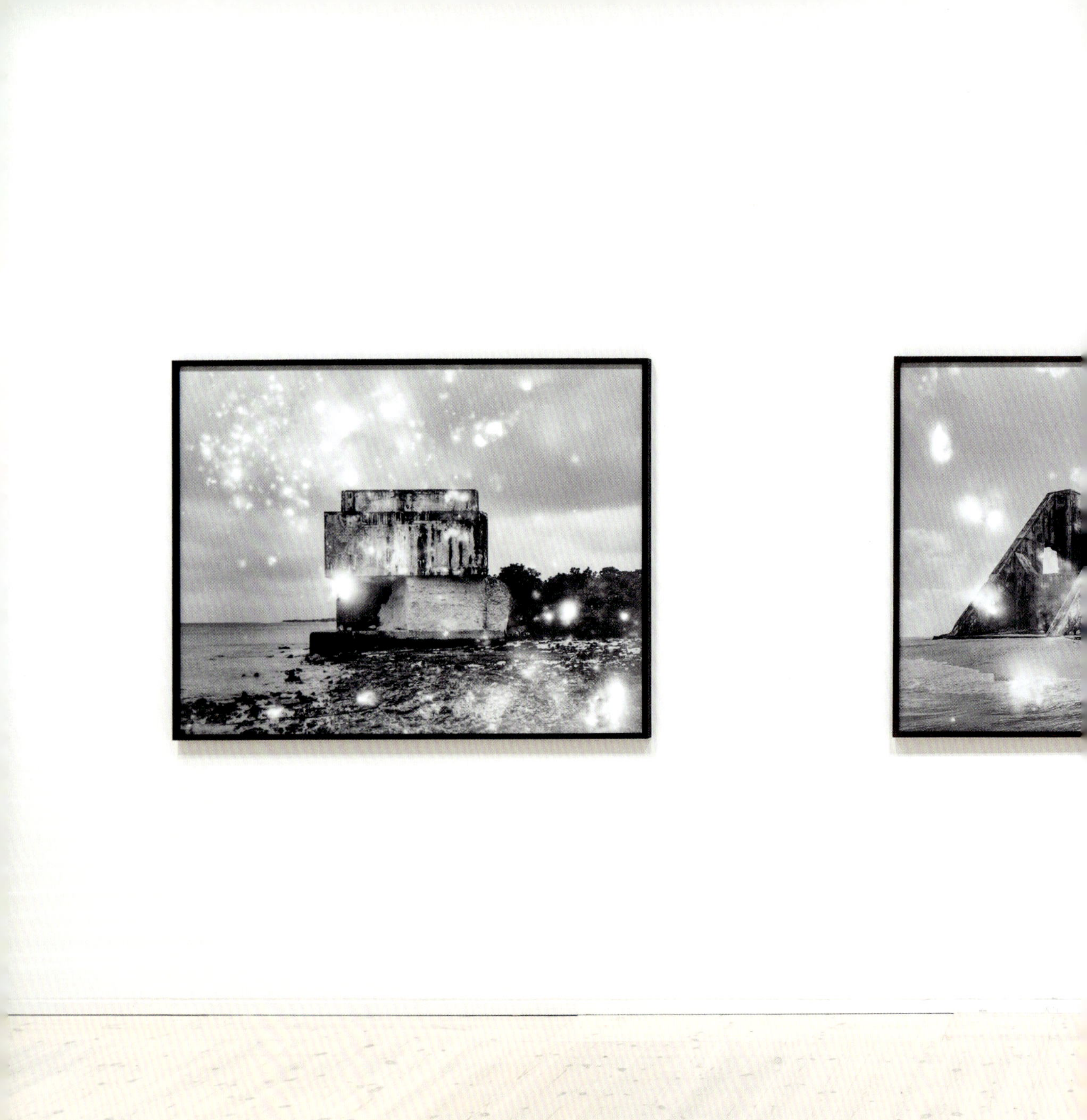

Terminal Beach
Julian Charrière
2016

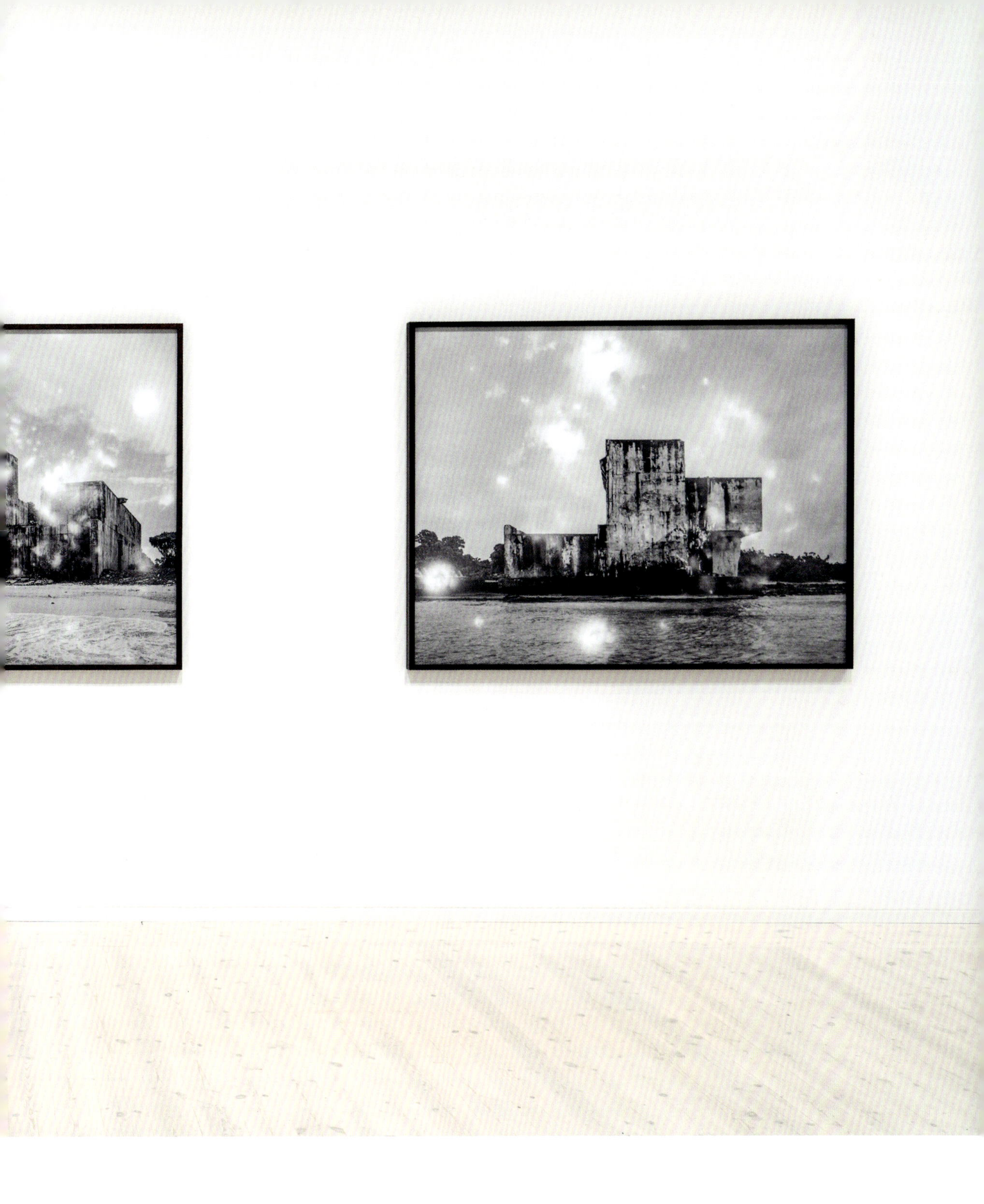

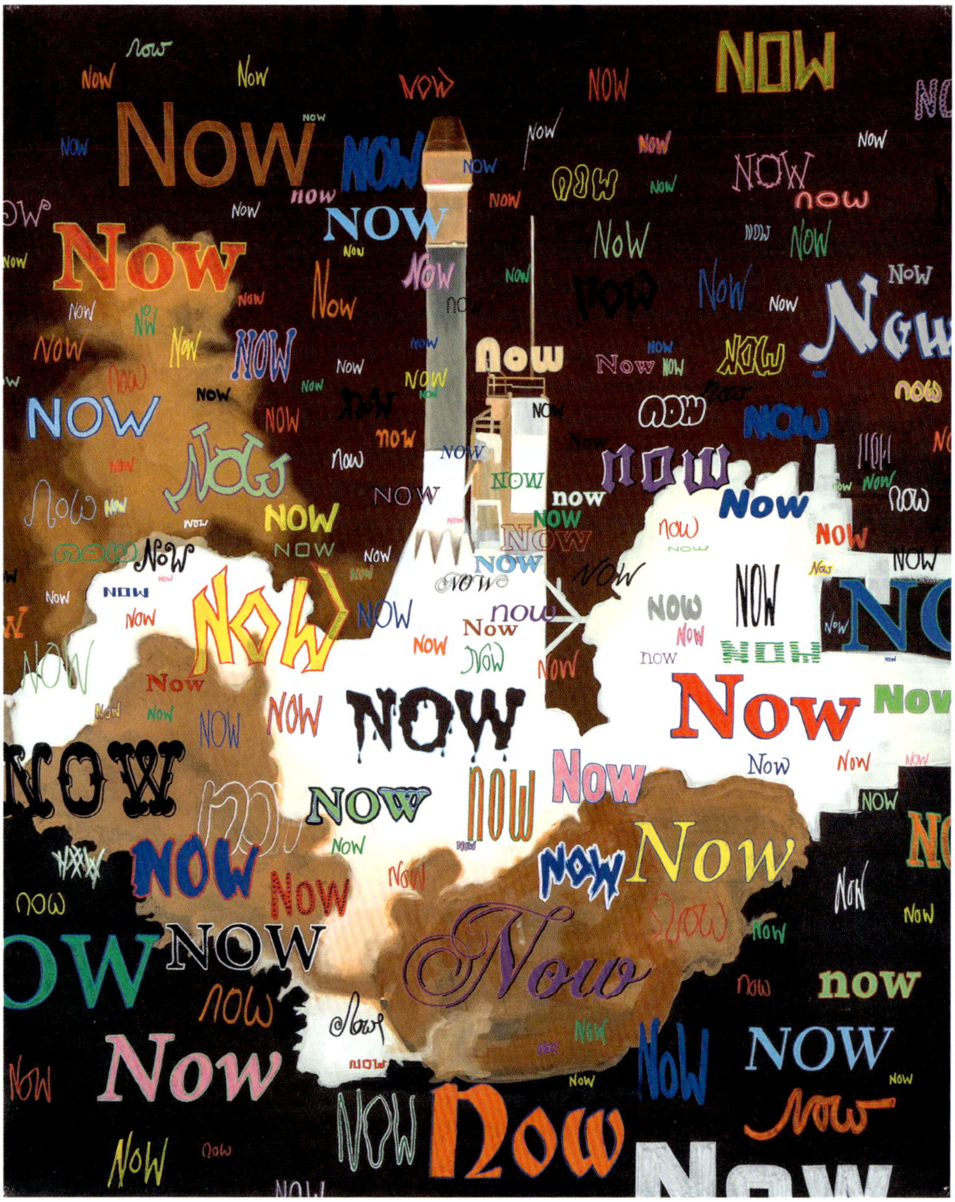

Now
Detail of installation
Selection of Studio Wall Drawings
Keith Tyson
1999–2017

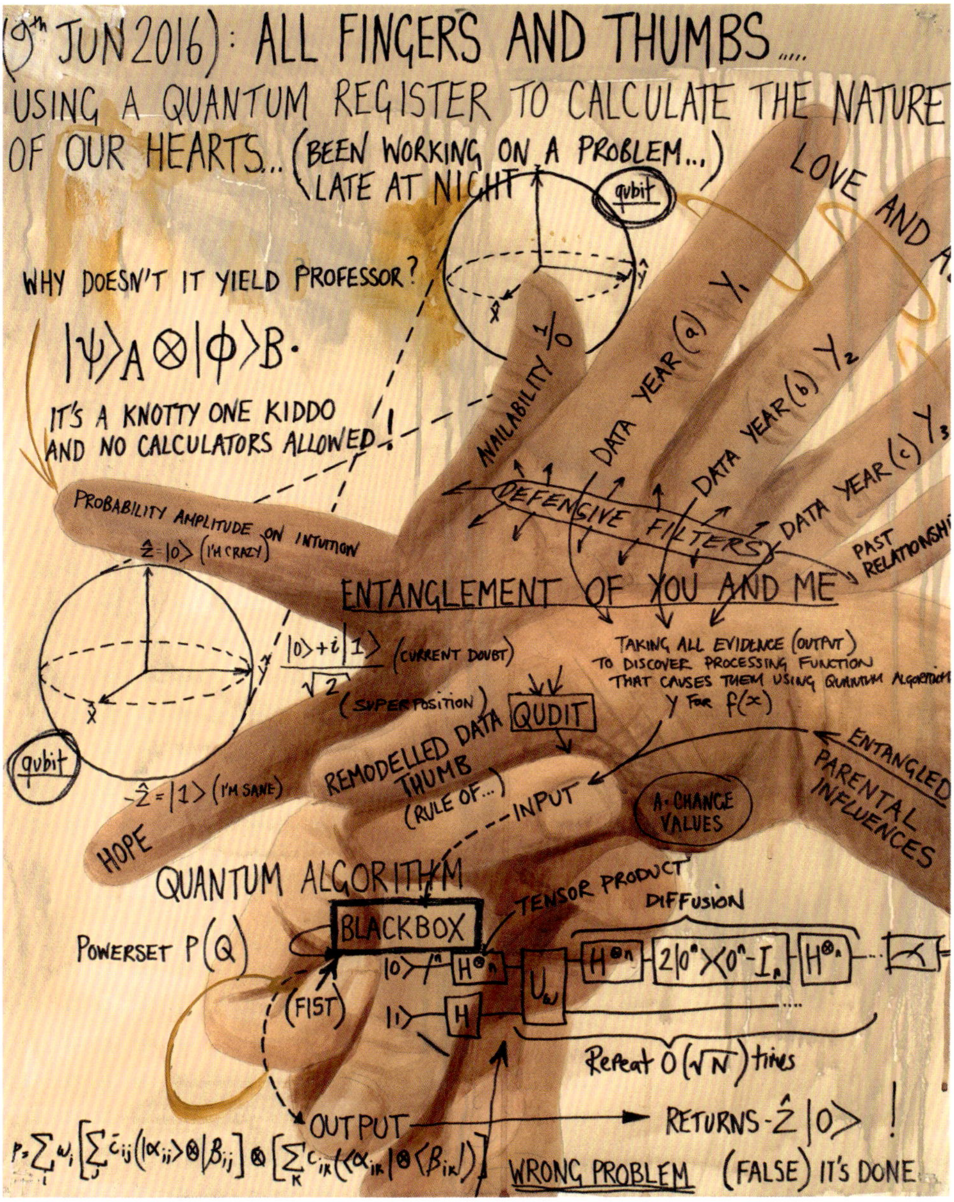

All fingers and thumbs
Detail of installation
Selection of Studio Wall Drawings
Keith Tyson
1999–2017

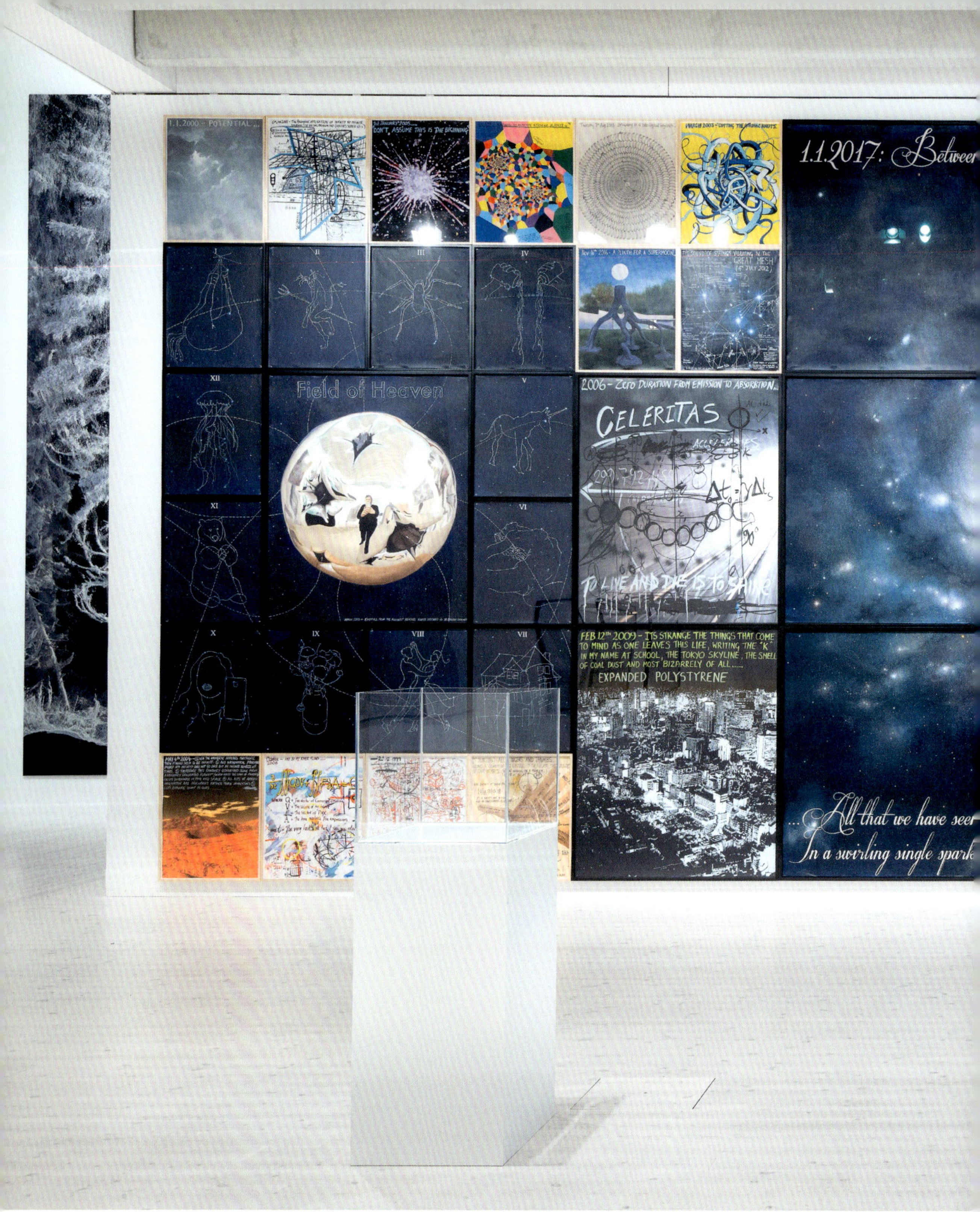

Selection of Studio Wall Drawings
Keith Tyson
1999–2017

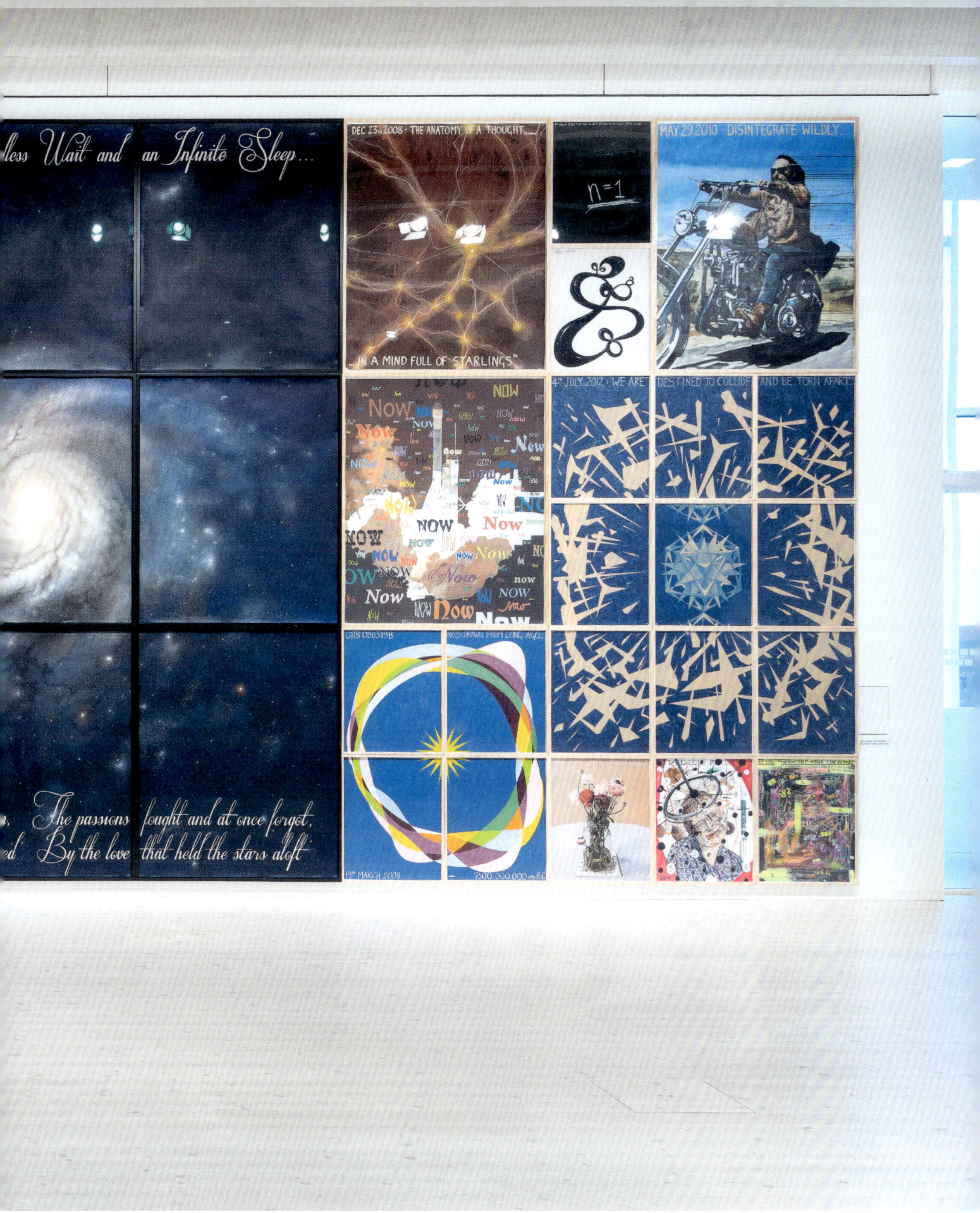
less Wait and an Infinite Sleep...
DEC 23rd 2008 · THE ANATOMY OF A THOUGHT
"... IN A MIND FULL OF STARLINGS"
n=1
MAY 29 2010 · DISINTEGRATE WILDLY...
4th JULY 2012 : WE ARE
DESTINED TO COLLIDE
AND BE TORN APART
NOW
GRB 080319B
ARCS DRAWN FROM LONG AGO...
19th MARCH 2008
7500,000,000 yrs B.C
The passions fought and at once forgot,
By the love that held the stars aloft

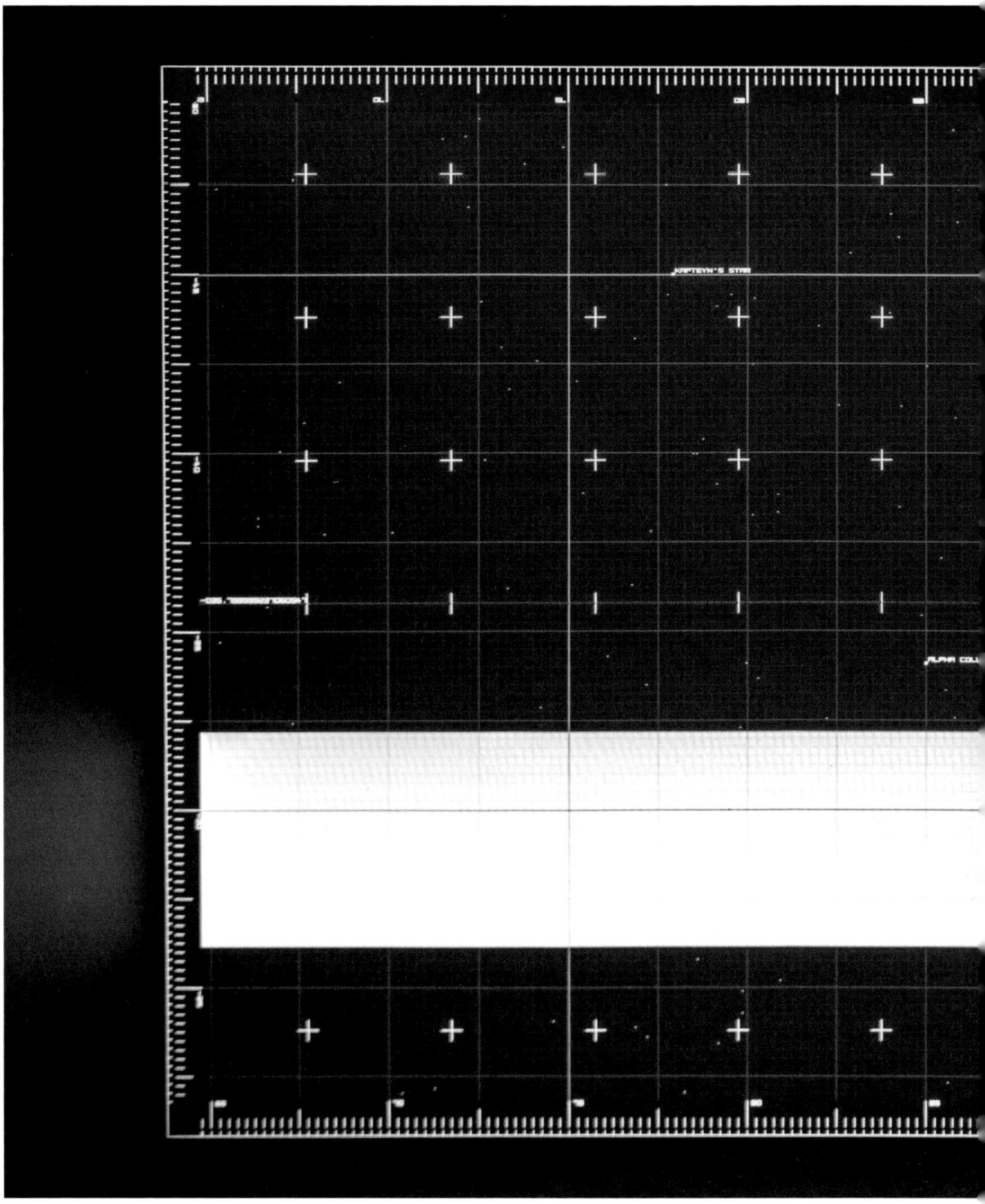

Data.tron (WUXGa version)
Ryoji Ikeda
2011–2012

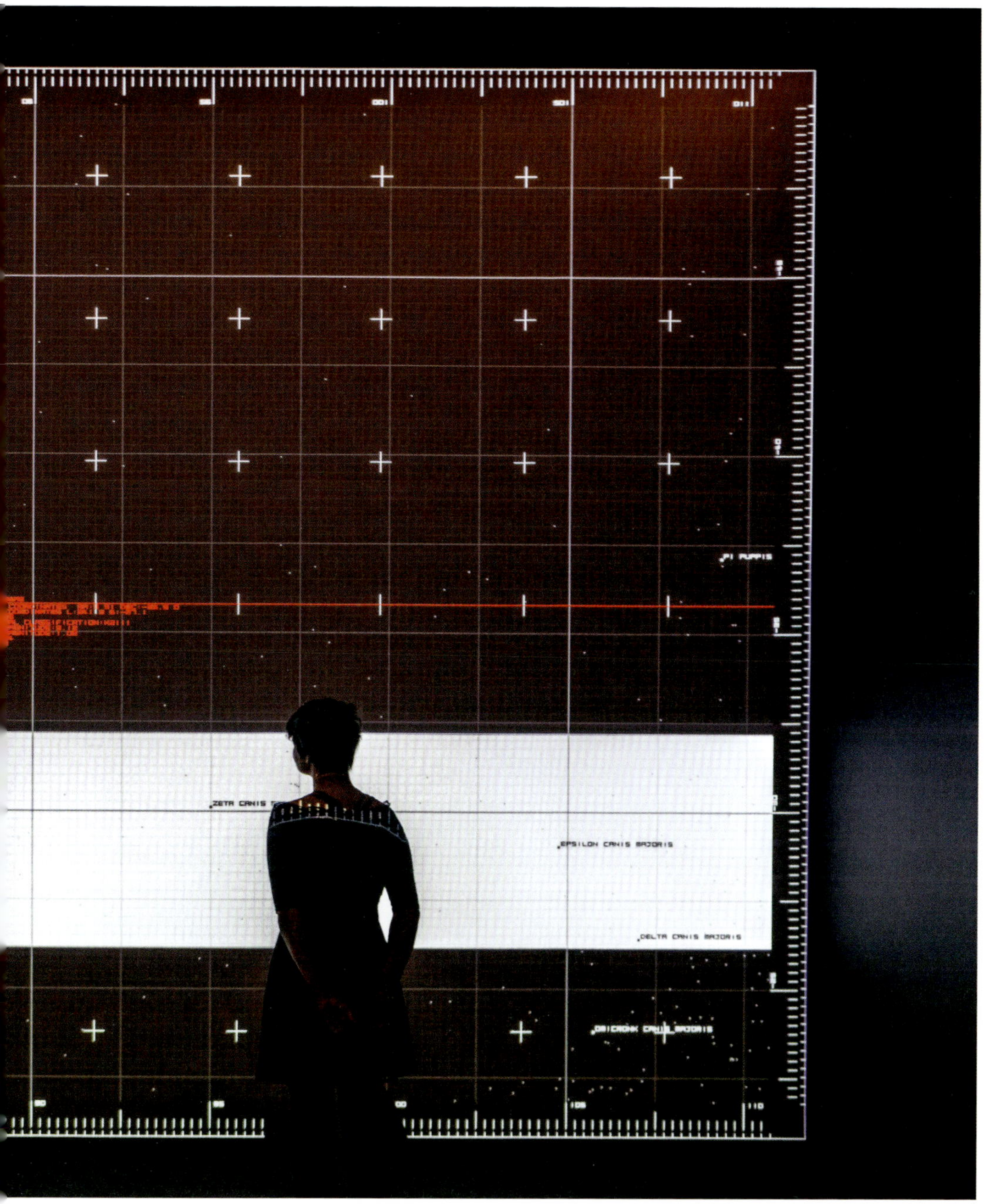
PI PUPPIS
ZETA CANIS
EPSILON CANIS MAJORIS
DELTA CANIS MAJORIS

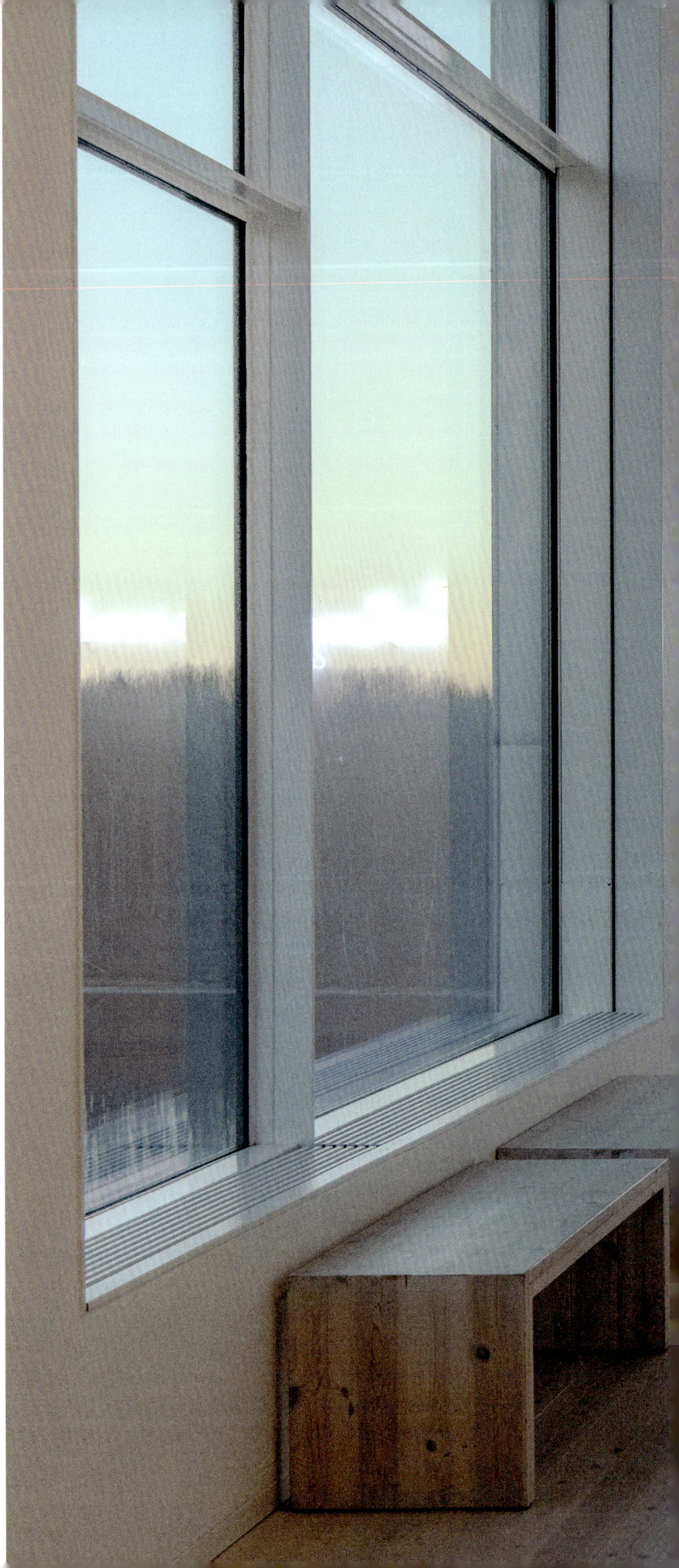

Entanglement
Rafael Lozano-Hemmer
2005

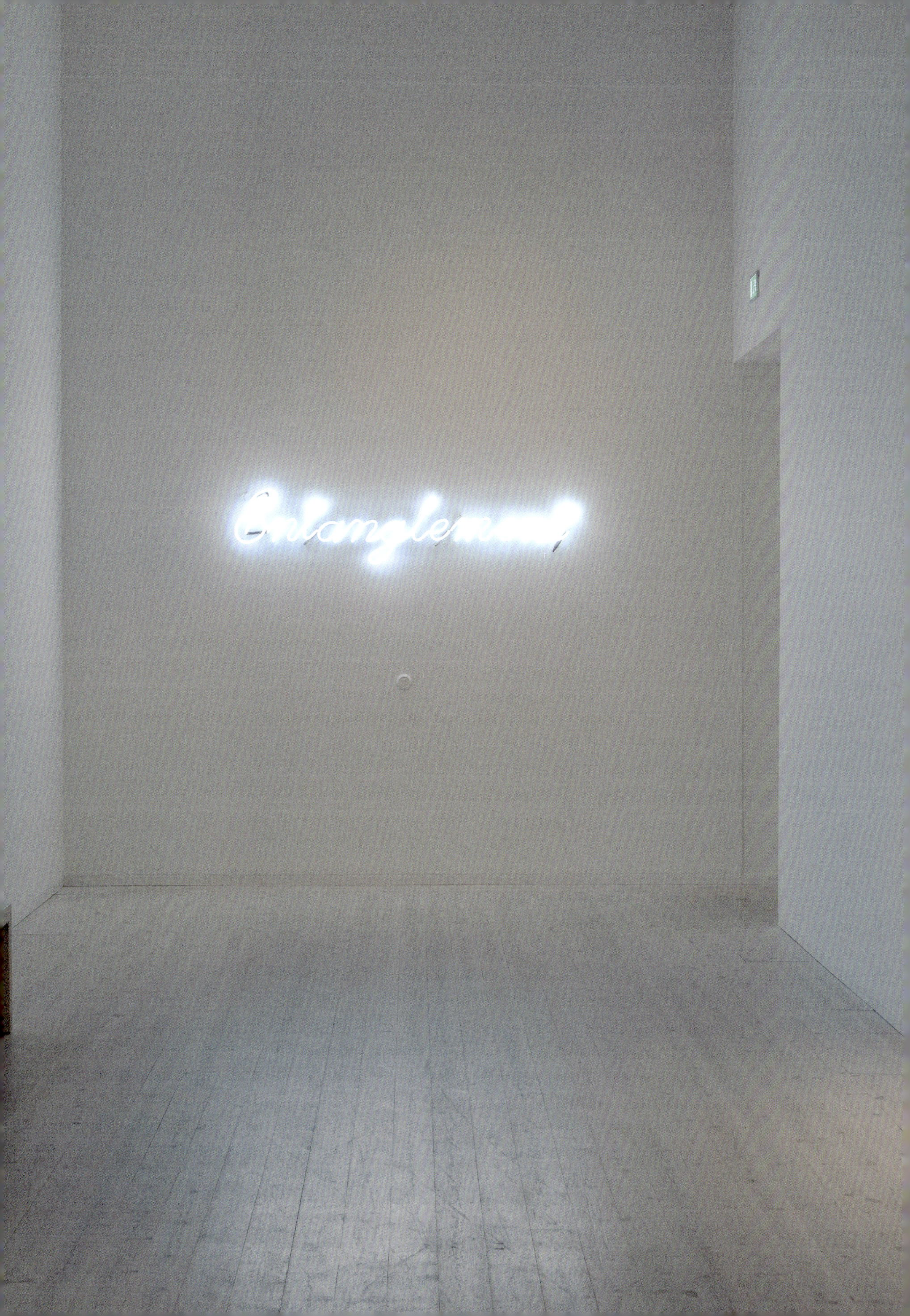

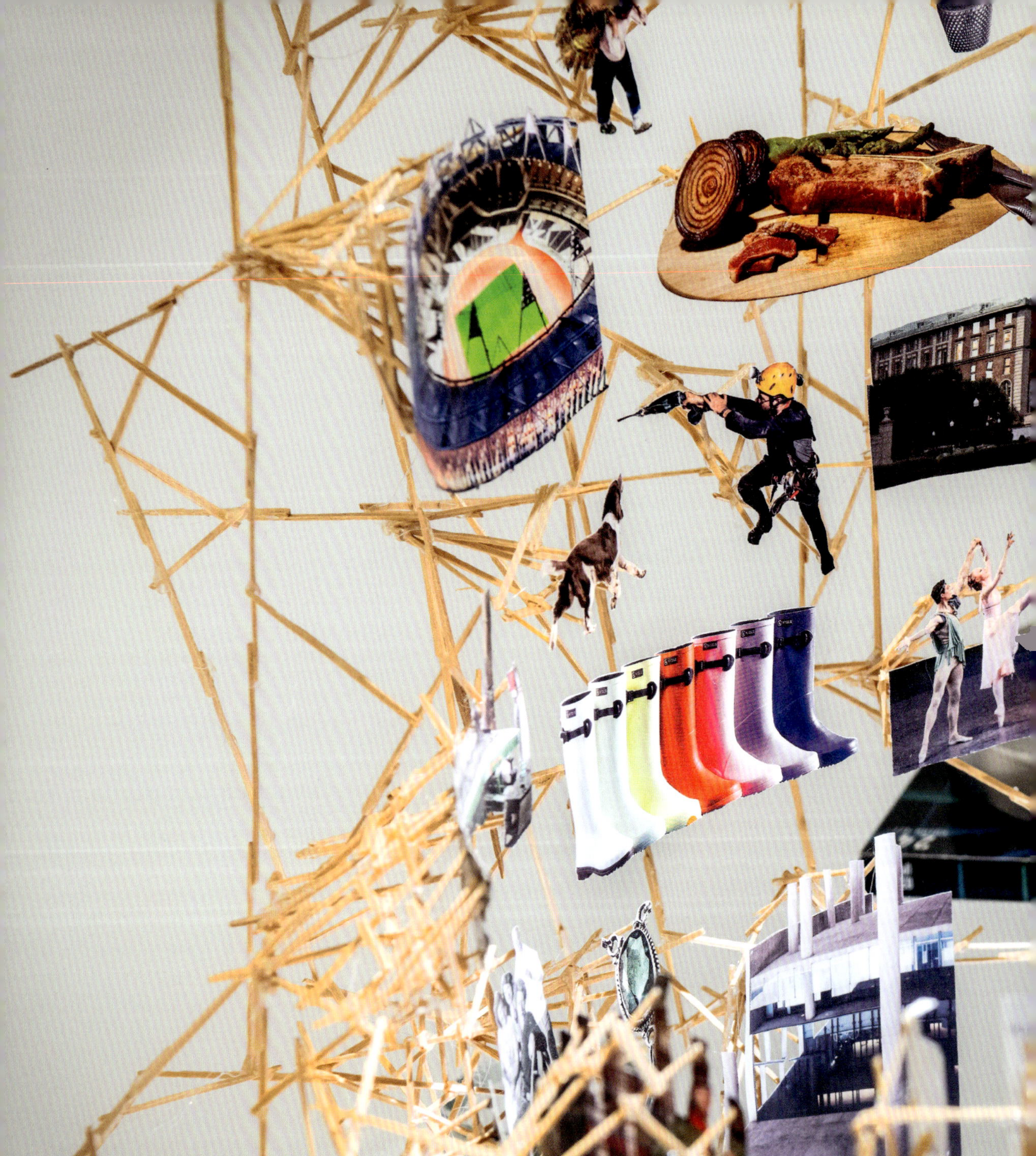

Detail
Dark Matter
Sarah Sze
2012

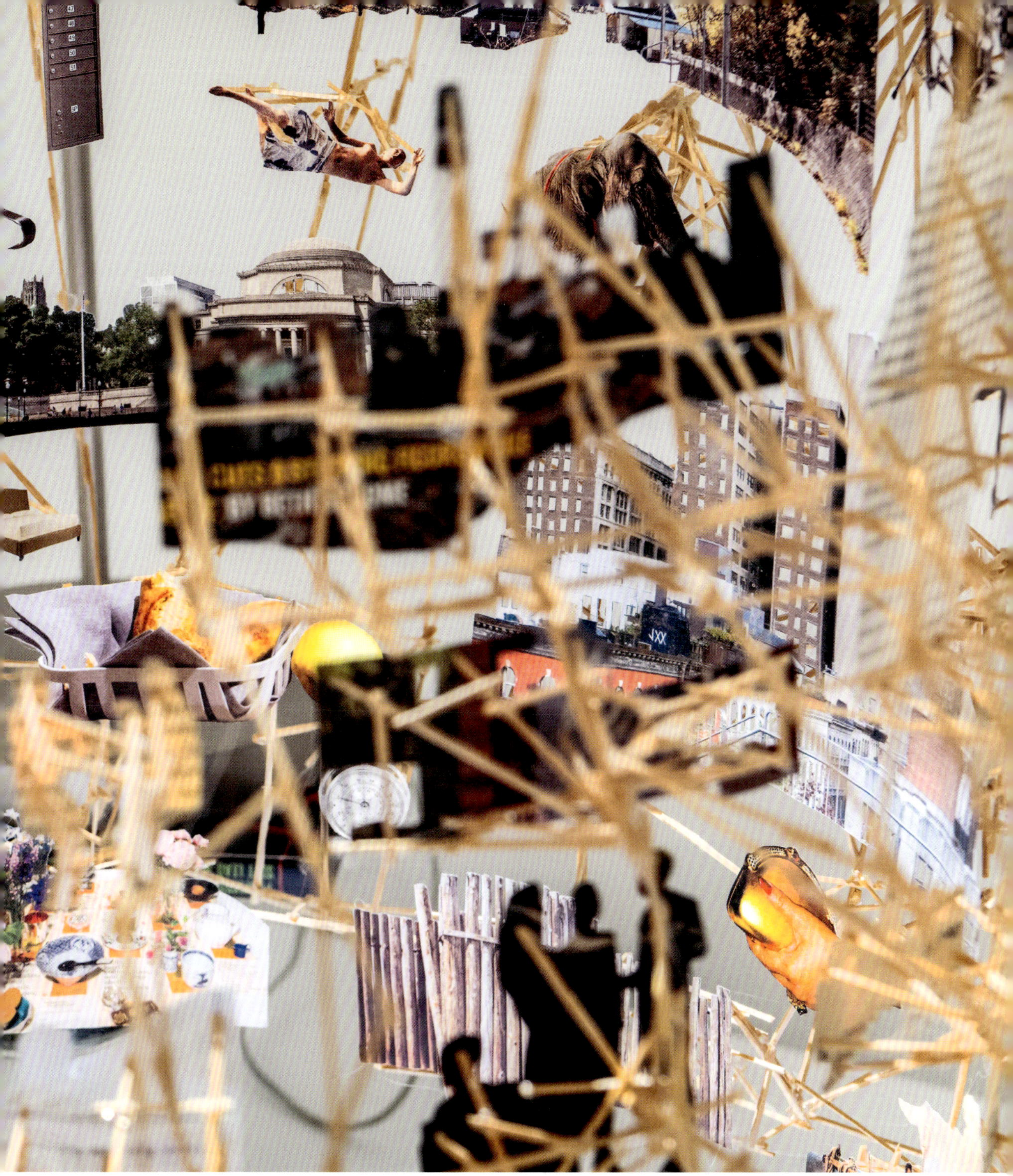

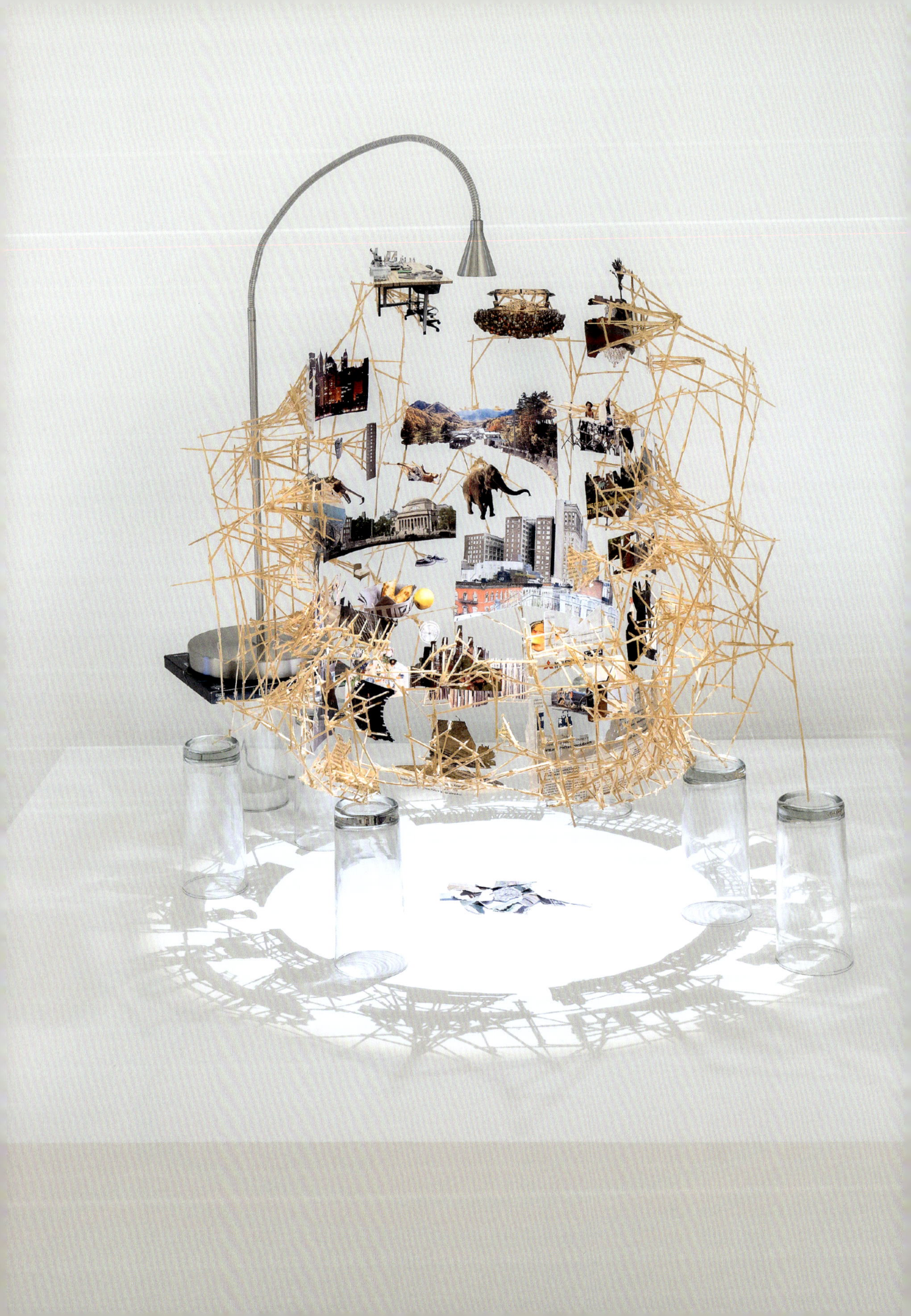

Dark Matter
Sarah Sze
2012

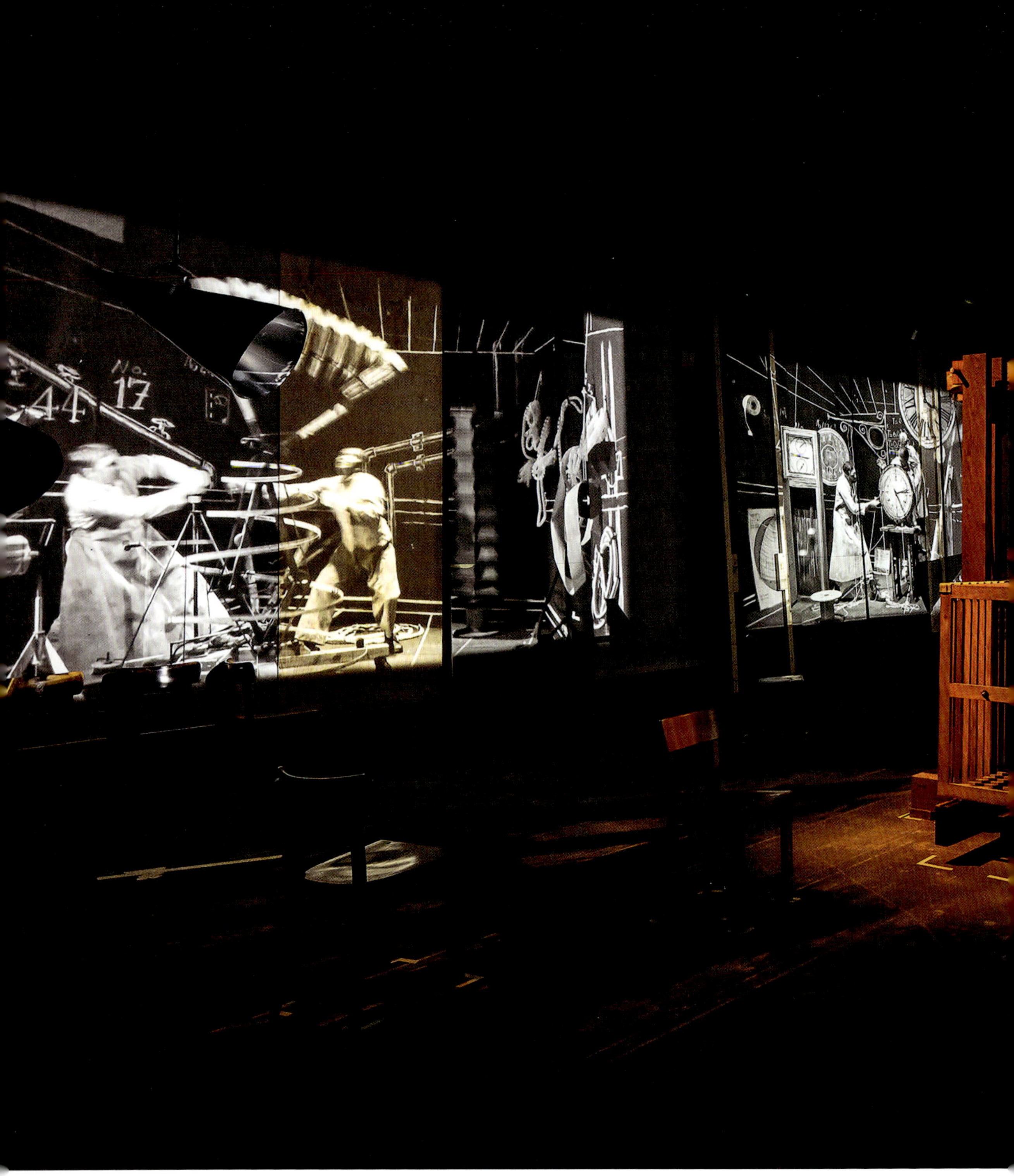

The Refusal of Time
William Kentridge
2012

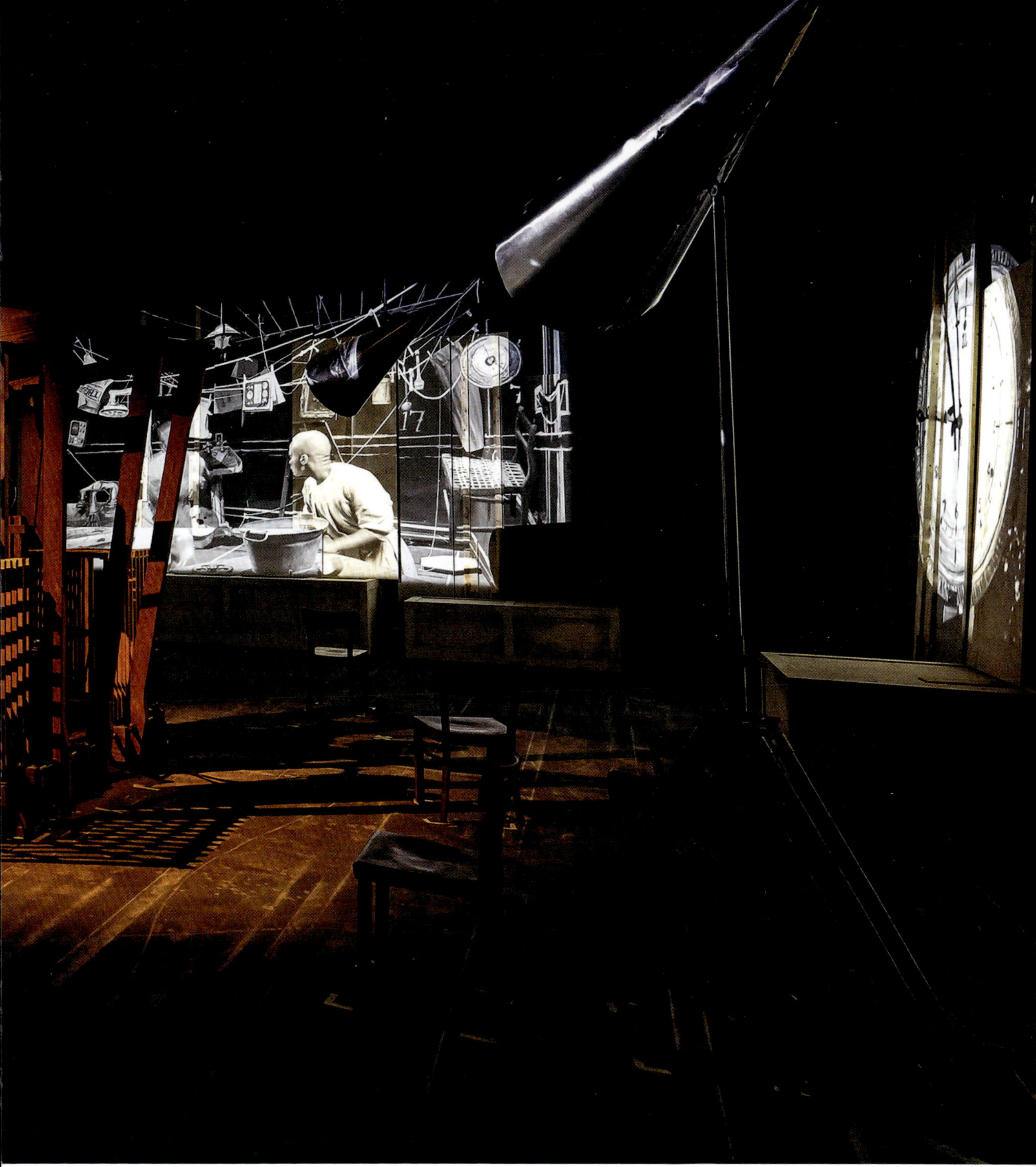

NOTES ON ARTWORKS IN THE EXHIBITION

Ariane Koek

Julius von Bismarck
***Two Self-Revolving Toruses*, 2010/2018**
(pp. 12, 94–95, 100)
Aluminum, Kevlar, polyester, motors
Courtesy of the artist

The shape of the universe has been postulated to be a torus. The *Self-Revolving Torus* in Julius's words represents an artistic and conceptual failure to comprehend let alone create an object which goes beyond three dimensions in our physical universe. In making the torus, he has created the illusion of it beyond those fixed earthly dimensions; and as it moves, it seems to be constantly consuming and reinventing itself by turning inside-out and outside-in, evoking a black hole.

Based in Berlin, Julius von Bismarck was the first Arts at CERN artist-in-residence on the Collide International program. He spent two months at CERN, the particle physics laboratory outside of Geneva. In his works, Julius von Bismarck explores people's perceptions and cultural preconceptions. He often uses the ideas and laws of physics to challenge the way we are used to seeing and knowing.

His interest in physics is from an early age. Julius von Bismarck's great-grandfather was a famous physicist who studied the sun and built solar observatories. He specialized in physics in high school, but decided to become an artist in the end, leaving him free to explore the idea of physics in his art. For example, at Art Basel in 2017, he spent days in a parabolic constantly spinning bowl, which challenged his body's perception of gravity and the way in which the human body interacts with it.

Julius von Bismarck
***Freedom Table and Democracy Chair*, 2016**
(pp. 98–99)
Two motors, computer, steel cable, office furniture from Bildmuseet
Courtesy of the artist

What does it mean when the objects we take for granted are no longer fixed on the Earth and, seemingly defying gravity, spin in space? What does this also do to meaning and the words we associate with things? This work by Julian von Bismarck is propelled by two motors which at certain times bring the table and chair into exact synchronicity with each other as they circle the air. They have a weightless freedom.

The title of the work is *Freedom Table and Democracy Chair*. With the political names given to these everyday objects, do they have any meaning at all when floating in space? Or is the piece in fact a critique of our ideas of freedom and democracy?

Julius von Bismarck
***Escape Shapes*, 2016 (pp. 96–97)**
Film assemblage (40 min.)
Courtesy of the artist

This film is a compilation of every major rocket disaster from the last seventy years shown in a forty-minute loop. Julius pulled footage from the NASA archives, from TV documentaries, and even from private individuals who had happened to capture the disasters.

Rocket science is a symbol of the competition between nations to go further or faster than each other. It is humankind's attempt to defy the forces of gravity and resist being earthbound. But in these explosions to go beyond the Earth's atmosphere, everything is lost, including, in two of these clips, human life. The piece is a comment on the fact that the advances of science and technology are built on failure. It is only by failing that one can learn and acquire further knowledge. And yet we never hear about this, let alone understand it.

Julius von Bismarck
***Lidar Helmet*, 2018 (new commission) (p. 115)**
Helmet, new media
Courtesy of the artist

Physics looks below the surface of nature and the world around us to show what lies beneath the surface and beyond our perception.

In this new commission, Julius von Bismarck invites us to look at the exhibition and spaces around us in a totally different way: to look beyond floors and walls, the materiality of the exhibits and the people moving through the exhibition. The *Lidar Helmet* enables us to see the world through new eyes thanks to a helmet which uses laser technology.

Like the way the human eye works with light in seeing objects, the lidar (which stands for light imaging, detection, and ranging) measures distance to an object by illuminating the target with pulsed laser light and measuring the reflected pulses with a sensor. Differences in laser return times and wavelengths can then be used to make digital 3D representations of the object.

Lidar's first applications were in meteorology, where the National Center for Atmospheric Research in the USA used it to measure clouds. The general public became aware of the accuracy and usefulness of lidar systems in 1971 during the Apollo 15 mission, when astronauts used a laser altimeter to map the surface of the moon.

Julian Charrière
***Terminal Beach*, 2016 (pp. 126–127)**
Black-and-white photograph, double-exposed through thermonuclear strata
Courtesy of the artist and Dittrich & Schlechtriem, Berlin

Known for his research-based practice rooted in geology, biology, physics, history, and archaeology, this French-Swiss artist frequently travels to some of the most remote regions of the planet to explore how human civilization and the natural landscape are inextricably linked.

The photographic series *Terminal Beach* was shot at the former atomic test site in the Pacific Ocean known as Bikini Atoll, where the artist conducted an expedition partially inspired by English novelist J. G. Ballard's short story *The Terminal Beach*.

The photographs were developed by scattering thermonuclear dust from the sites onto the negative, creating light spots which show the unseen forces which continually act on the landscape pictured. Thus, they both depict the site of nuclear radiation and bear the actual trace of the effect of radioactivity. They show the beauty and terror of light and decay.

This method of using radiation to develop the photographs recalls how in 1896 the French physicist Henri Becquerel discovered X-rays and radioactivity when storing photographic negatives and uranium salts in the same dark drawer. Exposing the film stock to radioactive material destroys one mode of visual information while at the same time adding another. The result is a doubly synthetic topography. The energy of the past infects the present and will continue to warp the future.

Julian Charrière
***Flathead - First Light*, 2016 (p. 16)**
Color photograph, double-exposed through radioactive material
Courtesy of the artist and Dittriech & Schlechtriem, Berlin

Julian Charrière's series of large-format photographic prints of typically idyllic tropical island scenes, *First Light*, explores the fraught interactions between industrial modernity and geography – addressing the atomic landscape and postcolonial ecology of Bikini Atoll.

The photographs are like holiday snapshots for a tourist brochure – depicting neon sunsets, beaches, and coconut trees. However, strata from nuclear "hot" sites have been placed on the large-format color negatives during their development process.

Thus, the photographs document and reenact the subliminal beauty present in the atrocities that shook the Marshall Islands during America's nuclear testing in the Cold War, whilst playing with modern ideas of relaxation and commodity culture. These prints oscillate between the peaceful cliché of tropical sunset photography and the destructive beauty issuing from atomic "second suns."

Sou Fujimoto
***Energy Forest*, 2016 (pp. 85, 86–87)**
Polycarbonate, acrylic, steel wires
Courtesy of the architect and MAXXI, National Museum of 21st Century Arts, Rome

With a photograph of Albert Einstein staring down godlike on his studio wall in Tokyo, it is clear that the Japanese architect Sou Fuijimoto has more than a passing interest in physics. In fact, he says that physics is the biggest influence on his architectural practice. At one point he even considered becoming a physicist.

His work often plays with concepts of inside-outside, up and down, creating new ways of conceiving space, breaking down notions of walls and form and other boundaries.

His moquette *Energy Forest* demonstrates this very clearly. With its spiraling form, which mimics nature, this representation of a space of the future is perforated and continuously moving with people, vehicles, and animals in a constant energy flow. The piece is actually a model for a futuristic energy fuelling station, showing the interaction between landscape, architecture, and nature and the spiraling exchange of energy between the them.

Iris van Herpen
***Magnetic Motion* (p. 64)**
Dress from the Spring/Summer collection 2015, in collaboration with the architect Philip Beesley

***Magnetic Motion* (p. 65)**
Shoes from the Spring/Summer collection 2015, in collaboration with the Dutch designer Jólan van der Wiel
Courtesy of the artist

With many international awards, the Dutch fashion designer Iris van Herpen is known for pushing couture, technology, and materials to their limits. She has said that she is "inspired by the impossible."

In 2013, Iris van Herpen was specially invited to the world's largest particle physics laboratory as part of the Visiting Artists strand of Arts at CERN. Her Spring/Summer 2015 Paris *Magnetic Motion* collection, including the dress in the exhibition and the accompanying shoes created by the designer Jólan van der Wiel, were directly inspired by the patterns of magnetic fields she learned about at CERN.

The Large Hadron Collider has a magnetic field 100,000 times stronger than the Earth's magnetic field. The shoes which accompany the dress were designed by using magnetic fields to form their shape and were then finished by hand. This was a way of Iris introducing chaos into her design process, using the forces of nature over which she has no control rather than technology as her tool for creating the shoes.

Iris van Herpen
Bubble dress from the *Seijaku* Autumn/ Winter collection 2016, in collaboration with the musician Kazuya Nagaya (pp. 13, 68) Courtesy of the artist

This dress features thousands of hand-blown glass bubbles coated with silicone which are stuck together to form a solid structure around the body. The bubbles appear weightless – and form a bioluminescent prism around the body. The piece evokes wishes to defy gravity and become airborne.

The collection was inspired by cymatics – the study of wave vibrations and frequencies – and the connection between the sound, vibrations, and physicality of matter. In cymatics, the higher the frequency of the sound wave, the more complex the visible patterns. What is specifically beautiful about the glass bubble dress is that the curved glass creates a magnification and a beautiful light lensing – like the water drops on insects which were the original inspiration for this piece. The dress also evokes the particles that make up the human body and all matter – including the concept that all matter is 95 percent empty. This is the first time this dress has been exhibited in a museum and gallery space.

Iris van Herpen
Two dresses from the *Aeriform* Spring/ Summer collection 2017, in collaboration with the architect Philip Beesley (pp. 66-67)
Courtesy of the artist

The *Aeriform* collection examines the nature and anatomy of air and airborne materiality and lightness. These two dresses comprise feathery-light metal lace geodesic garlands of flowers floating around the body like a silver cloud. They reflect and carry light through them.

The dresses were particularly inspired by a conversation about quantum foam between Iris van Herpen and Philip Beesley with the astrophysicist Subodh Patil, whom they met on one of their many visits to CERN. Iris and Philip were both intrigued by this idea in physics of tiny fluctuations in space-time where miniscule new dimensions unfurl and then furl back in on themselves. Thus, the dresses look as if they were in the process of appearing and disappearing at the same time.

Warren du Preez and Nick Thornton Jones
***Spatial Reverse*, 2015 (p. 11)**
Dance video
3 min., 40 sec
Courtesy of the artists

Visible and invisible forces, liquid darkness and dazzling light – all envelop a mysterious suspended form which is surrounded by a dress that is on the verge of not existing at all. Iris van Herpen was trained as a dancer, and the interaction between the movement of the body and the materials and her dresses is fundamental to her design. Since founding her eponymous label in 2007, she has collaborated with leading choreographers.

This collaboration is with the choreographer Russell Maliphant and the dancer Carys Staton. She is floating in space wearing the Particle Dress – a piece designed by Iris and inspired by the invisible movements and energies of particle collisions as, for example, in the Large Hadron Collider at CERN. The piece reminds us that humans are born out of and connected to the universe like babies suspended in amniotic fluid: after all, we are all – universe, matter, and human beings – made of stars and supernova.

The film was created for *MOVEment*, a unique project where fashion, dance, and film unite. It was commissioned by *AnOther Magazine*. Sound by Salvador Breed. Creative consultant Jerry Stafford.

Ryoji Ikeda
***Data.tron (WUXGa version)*, 2011–2012**
(pp. 132–133)
Mixed media installation
Courtesy of the artist

Japan's leading electronic composer and visual artist is fascinated by data, light, and sound. He shapes music, time, and space by mathematical methods and explores these phenomena as sensations, pulling apart their physical properties to reveal their relationships with human perception as well as our ways of knowing in the digital age.

Ikeda is inspired by quantum physics and by an exploration of mathematical concepts, data, and information. When in 2014 he won the Collide at CERN Residency Award, he spent time at the laboratory investigating quantum computing and data-gathering. His work is represented in the exhibition by *Data.tron* – a piece created before he visited CERN. This artwork reveals the vast and infinitely divisible substratum of information in the form of data – which are the smallest particles of encoded knowledge circulating in our technologically mediated world. This information flood is both his material and his subject matter. Deliberately, this work exceeds the limits of human rationality, intelligibility, and cognition in order to hint at the beautiful, mysterious, and troubling aspects of new technology.

William Kentridge
***The Refusal of Time*, 2012 (pp. 140–141)**
Installation
Courtesy of the artist

First exhibited at documenta 13 in Kassel, Germany, which had as its theme the arts and sciences, this piece was the result of conversations between the South African artist William Kentridge and the Harvard physicist Peter Galison. The extended series of discussions, which began in 2010 between Kentridge and Galison, touch on the history of the control of world time, relativity, black holes, and string theory.

The wooden "sculpture" at the center of this installation takes its inspiration partly from Peter Galison's revelation that, in Vienna and Paris during the nineteenth century, in an effort to coordinate all the clocks in the two cities, special underground pumping systems were created. The construction, which is nicknamed the "elephant," is a giant bellows-like structure or primitive breathing machine made of wood and sounding like a heartbeat or breath – evoking the idea of nature's timekeeping as well as humankind's attempts to control it with technology. Surrounding it are five screens which show scenarios inspired by the early silent era – such as those by the French film director Georges Méliès. William Kentridge's voice can be heard in the soundtrack as he explores different notions of time and mortality.

The Refusal of Time is a key work in the *Entangle* exhibition, showing how the ideas of physics and conversations with physicists can become springboards of the imagination.

Music and soundscape: Philip Miller. Video editing: Catherine Meyburgh. Dramaturge: Peter Galison.

Rafael Lozano-Hemmer
***Entanglement*, 2005 (pp. 134–135)**
Neon and light switch
Courtesy of the artist

The Mexican-Canadian artist Rafael Lozano-Hemmer makes poetic and imaginative interactive artwork using technology which he calls "the language of our time." His work ranges from large-scale public site works to small, intimate pieces. All involve the spectator in some way and for inspiration his work draws on different sciences, politics, sociology, poetry, music, and art history.

His piece *Entanglement* consists of two neon signs and ordinary light switches. It is displayed on two different floors of the museum and relates to the term used in quantum physics for the strange behavior of two particles that behave as one although separated in space.

The neon signs communicate with each other through signals turning them on or off. They never work independently of each other, even though they are in different places – a commentary on Einstein's description of entanglement as spooky action at a distance.

Goshka Macuga
***International Institute of Intellectual Cooperation, Configuration 5 & 7*, 2015–2016 (pp. 78–79, 80)**
Bronze
Courtesy of the artist

Systems of knowledge, how they are classified and gain power in society, history and the illusion of objectivity – all are a source of endless fascination for the Poland-born artist Goshka Macuga. Taking on the role as artist, curator, researcher, historian, and designer, Macuga works in many different modes as well as with different media. She deliberately challenges herself and her artistic practice by not being fixed.

The title of her series *International Institute of Intellectual Cooperation* refers to a historical gathering of scientists and scholars that included the physicists Albert Einstein and Marie Curie. They met in the 1920s and 1930s as a council of elders for the League of Nations. She uses rods to interconnect specially made bronze heads of notable scientists and thinkers from history who took an interdisciplinary approach. Beyond the physical constraints of time, space, and history, in this installation they encounter each other as a constellation of thinkers in the molecular structure that she has created. Thus, in these hypothetical gatherings, there are potential new conversations and links to be made.

Configuration 5 is entitled *Beginning* and includes Charles Darwin (1809–1882), the English naturalist, biologist, and geologist known for his contribution to the science of evolution, Madame Blavatsky (1831–1891), the Russian occultist and co-founder of the Theosophical Society, the astronomer Nicolaus Copernicus (1473–1543), who was born in Poland and who formulated the model of the sun being at the center of the universe, the English radio astronomer Martin Ryle (1918–1984), the mathematician and astronomer Fred Hoyle (1915–2001), who rejected the idea of the Big Bang, the Russian physicist and creator of inflationary cosmology Andrei Linde (b. 1948), the English physicist and cosmologist Stephen Hawking (1942–2018), the Soviet cosmonaut and pilot Yuri Gagarin (1934–1968), the first person to create a cosmological point of view of the world, the Greek philosopher Anaximander (610 BCE–c. 546 BCE), and the Italian theoretical physicist and international best-selling science writer Carlo Rovelli (b. 1956).

Configuration 7 is called *Transhumanism* and includes the German art historian and cultural theorist Aby Warburg (1866–1929), the writer of *Frankenstein*, Mary Shelley (1797–1851), the Monster of Frankenstein (1818), the futurist and computer scientist Ray Kurzweil (b. 1948), who invented the concept of singularity, the American science writer Ronald Bailey (b. 1953), and the political scientist and author of *The End of History and the Last Man*, Francis Fukuyama (b. 1952).

Goshka Macuga
***International Institute of Intellectual Cooperation, Configuration 7 (continued)*, 2015–2016 (p. 16)**
Bronze
Courtesy of the artist

The conversation and connection between leading thinkers continue to seemingly break through the ceiling from the gallery below into the gallery above. New thinking and ideas know no bounds or boundaries. There are connections between the past, present, and future – all are deeply entangled. The

American feminist thinker and writer of the *Cyborg Manifesto* Donna Haraway (b. 1944) joins the Italian Renaissance nobleman and humanist philosopher Giovanni Pico della Mirandola (1463–1494) and the British evolutionary biologist, eugenicist, and internationalist Julian Huxley (1887–1975).

Goshka Macuga
***Untitled*, 2018 (new commission)**
(pp. 14–15)
Various blackboards, chalk
Courtesy of the artist

Goshka Macuga was a Visiting Artist to CERN in 2012 and was fascinated by how, in our highly advanced technological age, physicists use the most basic tools – blackboards and chalk – to work out their theories of our universe.

This ephemeral way of working out and marking new knowledge inspired her to create this new commission. Working with readymades as well as commissioning new work, this piece is all about knowledge-making – and how, like chalk on the blackboard, unless preserved behind glass, it is always wiped out and replaced by new ways of knowing. Knowledge is always evolving and building on the dust of the past.

In this installation, which she curated herself, Goshka Macuga includes seven colorful blackboard drawings by Rudolf Steiner, the founder of anthroposophy. He resorted to using colored chalk to illustrate his most complex thoughts when he was not able to communicate in words.

There is a blackboard by Joseph Beuys called *Continuum* showing time's arrow, and nine newly made ones inscribed with the favorite equations of nine contemporary physicists, including Carlo Rovelli, who wrote a text for this catalogue and is also represented in Goshka Macuga's *Institute of Intellectual Cooperation*, and four leading scientists who work at CERN, Professor Michael Bradley from the Department of Physics at Umeå University, and Cecilia Jarlskog, Lund University.

Also included in the installation is one of the two blackboards said to have been used by Albert Einstein at Oxford University where he lectured in May 1931. This is the first time that this blackboard has traveled outside of the UK.

Provocatively, the Einstein blackboard included in the installation is not the famous one with the equation of a model of the universe proposed by Einstein in May 1931. Instead, it is the lesser known blackboard which was said to have been wiped clean by an overzealous cleaner, although traces of the chalk he used can be seen. This blackboard shows that our knowledge of the universe can be wiped clean at any moment, with traces of what we knew before being left behind, ready for new ideas to fill their place.

Joseph Beuys
***Continuum*, 1978–1984 (pp. 14, 77)**
Chalk on blackboard
Courtesy of Galerie Thaddaeus Ropac, Paris

Rudolf Steiner
***Untitled*, 1921–1924 (pp. 15, 74–75)**
Colored chalk on black paper
Courtesy of the Rudolf Steiner Archive, Dornach

Blackboard drawing from lecture, Jul. 15, 1921
Blackboard drawing from lecture, Oct. 9, 1921
Blackboard drawing from lecture, Jan. 2, 1922
Blackboard drawing from lecture, Oct. 31, 1923
Blackboard drawing from lecture, Dec. 21, 1923
Blackboard drawing from lecture, Apr. 19, 1924

Albert Einstein
Blackboard with white chalk traces, 1931
(pp. 15, 77)
Courtesy of the History of Science Museum, University of Oxford, Inv. 11714

Used on May 16, 1931, during his lectures at the University of Oxford, United Kingdom.

Blackboards written with equations, 2018 (pp. 14, 76)
Courtesy of the authors:

Michael Bradley, physicist, Umeå University, Sweden; Bilge Demirköz, astroparticle physicist, Ankara University, Turkey, and the AMS experiment, CERN; Michael Doser, experimental physicist at CERN and spokesperson for the Aegis experiment; John Ellis, theoretical physicist at CERN; Subodh Patil, theoretical cosmologist, Niels Bohr Institute, University of Copenhagen, Denmark; Hiranya Peiris, astrophysicist at University College, London, UK; Carlo Rovelli, theoretical physicist, Aix-Marseille University, France; Cecilia Jarlskog, theoretical physicist at Lund University, Sweden; and Peter West, theoretical physicist at King's College in London, UK.

Davide Quayola
***Remains*, 2018–ongoing (pp. 6–7, 112)**
3D scan, digital print
Courtesy of the artist and Audemars Piguet

Davide Quayola investigates the unpredictable collisions, tensions, and equilibriums between the real and the artificial, the figurative and the abstract, the old and the new. His work explores photography, geometry, time-based digital sculptures, and immersive audiovisual installations and performances.

Remains is part of Quayola's ongoing focus on nature and the tradition of landscape painting. While re-creating similar conditions to en plein air painters of the late-nineteenth-century, the natural landscapes are actually observed and analyzed through the "eye of machines" to reveal another world.

Quayola captured the trees in this piece using drones with high-precision laser scanning technology developed by physics. The result materializes as an ultra-high-resolution large-format print. It is one of a large series of pieces with the same title created in this way.

From a distance, the piece looks as if it were just a photograph of trees. Moving closer, one realizes that it is a computer-generated painting made up of billions of points. It is a tree formed out of data bytes, and thus the piece comments on the way we now mediate nature in our technological world, whilst also ironically evoking the invisible particles which make up our natural world.

Solveig Settemsdal
***Singularity*, 2016 (p. 106 and cover image)**
Video, 9:27 min.
Courtesy of the artist

The effects of pressure, heat, gravity, and light are integral to this work by the Norwegian-born artist Solveig Settemsdal. She investigates the transience and fluidity of materials and is inspired by the ideas of science, including chemistry and physics.

Taking its title from the term in physics for how a small change can cause a large effect, *Singularity* explores drawing as both a temporal and sculptural process which evolves like a living organism. A point appears in a perceived void. Slowly expanding, its articulation grows increasingly deliberate; lines are created, crossed, and bisected in a fluid three-dimensional space by suspending white ink in gelatin. Randomness finds form.

In 2016, *Singularity* was awarded first place in the Jerwood Drawing Prize. The piece was created in collaboration with the musician Kathy Hinde, who composed the sound score. It begins by grouping initially abstract sounds, from the violin and prepared toy pianos, and becomes an increasingly organized soundtrack.

Sarah Sze
***Dark Matter*, 2012 (pp. 136-137, 138)**
Mixed media
Courtesy of the artist and Tanya Bonakdar Gallery, New York/Los Angeles

Sarah Sze's work is in a constant state of becoming. She reveals the process of making and creation by exploding or imploding structures made out of everyday things – Q-tips, found objects, wire, photographs, sticky tape. She creates small universes out of the debris and detritus of life, showing how the tiny and discarded hold an infinity of possibilities. What appears to be random is actually very precise.

Playing with the substantiality of things and the world around us, she shows how our memories and identity are in a state of constant flux, like the universe itself. Her mixed media work *Dark Matter* comprises upturned glasses, photographs, and postcards. Its title is taken from the physics term for a hypothetical form of matter which we cannot see and which is thought to comprise 85 percent of all matter in the universe.

It is a piece which evokes the idea that our worlds, including our memories and experiences, are made up of fragments which together make a whole. It is also about that most invisible of processes – creativity – and the putting together of fragments which make a finished artistic piece whole.

Keith Tyson
***Selection of Studio Wall Drawings*, 1999-2017 (pp. 128-129, 130-131)**
Paintings, photographs, collage
Courtesy of the artist

The question of how and why things come into existence is a thread throughout the career of the British artist and Turner Prize winner Keith Tyson. He is fascinated by natural structures as well as the attempts by science, including physics, to understand them. His 2002 Supercollider at South London Gallery was inspired by CERN science and particle physics, even before he actually visited the laboratory in 2009.

In our exhibition, he creates a new kind of supercollider wall, comprising assembled paintings, photographs, and collages from his studio which investigate different physical phenomena. He makes these new collisions between seemingly different media, modes, and perceptions across different times and spaces of making. For the artist, these pieces have a personal intra-connection and entanglement.

Jorinde Voigt
***Space Study 1-4*, 2016**
Ink, gold leaf, oil pastel, pencil
***The Blue Shift I*, 2017**
Ink, gold leaf, pastel, chalk, drawing (pp. 94-95, 122-123, 124-125)
Courtesy of the artist

Originally trained as a photographer, the Berlin-based artist Jorinde Voigt turned to drawing and painting as her medium in order to investigate the imperceptible and invisible. Her works have been likened to musical scores, scientific diagrams, or notational thought models. Using a precisely coded system of mark-making, she gives pictorial form to an array of natural and psychological phenomena, inspired by the ideas of philosophy, science, music, and literature, as well as by her personal experience of the world.

Space Study 1-4 shows how drawing for Jorinde Voigt is an exploration of time and space – whether geographically, emotionally, or physically. The body and perceptions are given tactile engagement with the world by the act of drawing and making notes by hand. In *The Blue Shift I*, the position of the artist's body is a fundamental part of the work. What at first looks like organs or amorphous shapes turn out to be the contact points of parts of her body on paper. The title of the piece is from a key concept in physics. The "blue shift" is a term in physics used to describe any decrease in wavelength with a corresponding increase in frequency of an electromagnetic wave if the sender moves toward you. In visible light, this shifts the color from the red end of the spectrum toward the blue end. The result is that if objects such as stars or galaxies approach you, their color becomes more blue. As the title of the piece indicates, perception shifts and changes. In this case it is the artist's body in relation to the painting, with the body becoming part of the work and painting process.

Carey Young
***Missing Mass*, 2010 (p. 120)**
5,461 dark matter particles present in open-top Perspex container, on a plinth with text
Courtesy of the artist and Paula Cooper Gallery, New York

Missing Mass is a work which deliberately plays with the disinterested language of physics and with how scientific knowledge is always changing. A transparent Perspex box has stenciled on its pedestal the explanation that the box holds 5,461 particles of the theoretical substance known as dark matter which is said to make up 85 percent of all matter in the universe. It is transparent and undetectable to the human eye. The text explains that the particles may circulate freely inside the box, their number may change, and all this information is provisional, "based on current scientific understanding."

By treating this hypothetical substance as a readymade, Carey Young provocatively links theory and the imagination to appropriation and control.

Scientific advice for this piece was provided by the physicist Dr. Malcolm Fairbairn, King's College London.

Carey Young
***Contracting Universe*, 2010 (p. 119)**
Digital image on adhesive vinyl, white bench
Courtesy of the artist and Paula Cooper Gallery, New York

NASA's recent digital rendering of the surface of Mars is blown up in this work by Carey Young to closely resemble a modernist photograph of a sublime mountainscape. When viewed at close range, the image appears more and more like a digital approximation, suggesting a poetic futility inherent in the human desire to comprehend or manage the infinite.

The title refers to the theory in physics that the ultimate fate of the universe might be that space contracts – a theory also known as the Big Crunch. In this theory, the universe might contract so much that it could ultimately end in either a black hole or might lead to another Big Bang in which the earth expands and re-forms again.

Carey Young
***Report of the Legal Subcommittee*, 2010 (p. 121)**
Inkjet print
Courtesy the artist and Paula Cooper Gallery, New York

Carey Young has developed her artistic practice working across the disciplines of science, economics, politics, communications, and law. This print by Carey Young features a map of the stars, together with a found transcription of a United Nations meeting in which international delegations declare frustration with their forty-year-old, ongoing efforts to devise a legal definition of outer space.

This piece shows how the human attempts to bureaucratize and control outer space, including the stars, are so far seemingly frustrated by the sublime scale and the mystery of its infinite depth.

NOTES ON ARTISTS, PHYSICISTS, AND WRITERS

PHILIP BALL is a popular freelance science writer. Philip is the author of many popular books on science, including works on the nature of water, pattern formation in the natural world, color in art, the science of social and political philosophy, the cognition of music, and physics in Nazi Germany. Philip writes regularly for *Nature*.

JULIUS VON BISMARCK is a German artist currently living and working in Berlin, Germany. He studied and worked in New York and Berlin and completed his studies with Olafur Eliasson at the Institut für Raumexperimente in Berlin. Bismarck was the first Collide at CERN artist in residence in 2011 held in conjunction with Ars Electronica.

MICHAEL BRADLEY is professor of physics at Umeå University, Sweden, working in general theory of relativity.

JULIAN CHARRIÈRE is a Swiss conceptual artist currently living and working in Berlin, Germany. He uses a range of artistic approaches, including photography, performance, sculpture, and video to address concepts relating to time, nature, and society's relationship to the natural world. Charrière was a former pupil of Olafur Eliasson's, where he met Julius von Bismarck, with whom he shares a studio and often collaborates.

BILGE DEMIRKÖZ is an astroparticle physicist and professor of high energy physics at the Middle East Technical University, Ankara, Turkey. She is also part of the AMS (Alpha-Magnetic Spectrometer) collaboration on the International Space Station. Since 2014 she has served on the board of the Arts at CERN program.

MICHAEL DOSER is a senior research physicist at CERN and is spokesperson of the AEgIS experiment. He specializes in working with antimatter, using it either as a tool or as an object of study itself. He was a founding member of the board for the Arts at CERN program.

JOHN ELLIS is a British theoretical physicist who has worked for four decades at CERN. He is currently the Clerk Maxwell Professor of Theoretical Physics at King's College London, UK. His research interests are in particle physics, astrophysics, cosmology, and quantum gravity.

SOU FUJIMOTO is one of Japan's most creative and visionary architects, with offices in Tokyo and Paris. After he graduated from the Department of Architecture, Faculty of Engineering, at Tokyo University, he established Sou Fujimoto Architects in 2000.

PETER GALISON is the Joseph Pellegrino University Professor of the history of science and physics at Harvard University. His numerous publications include *Image and Logic: A Material Culture of Microphysics* (1997) and *Einstein's Clocks, Poincaré's Maps: Empires of Time*. In 2010, he began collaborating with the artist William Kentridge, which led to the creation of what is considered William Kentridge's masterwork – *The Refusal of Time*.

IRIS VAN HERPEN is an award-winning fashion designer based in Amsterdam. Nature and the latest developments in science and technology are the inspirations for her futuristic work, and she works in close collaboration with diverse practitioners. In 2014, she was a Visiting Artist at the Arts at CERN program and created the Spring/Summer 2015 collection *Magnetic Motion* as a result.

RYOJI IKEDA is a Japanese visual and sound artist living in Paris. He focuses on the essential characteristics of sound itself and of visuals as light. His work includes immersive live performances and installations. In 2014, he was awarded the third Collide International artist-in-residence at CERN held in conjunction with Ars Electronica.

CECILIA JARLSKOG is Professor Emerita of mathematical physics at Lund University, Sweden. She is mainly known for her study and expertise in theoretical particle physics. Her studies include research on the ways that subatomic and electronic constituents of matter cohere or lose their symmetry and on grand unified theory.

WILLIAM KENTRIDGE is a South African artist best known for his prints, drawings, and animated films. They are constructed by filming a drawing, making erasures and changes, and filming it again. These palimpsest-like drawings are later displayed along with the films as finished pieces of art, opera sets, and installations.

ARIANE KOEK is an acknowledged international producer and pioneer in the field of arts, science, and technology. She initiated, designed, and directed the acclaimed Arts at CERN program from 2009 to 2015. Koek consults, writes, and lectures extensively about arts, science, and technology, residency programs and interdisciplinary creativity.

RAFAEL LOZANO-HEMMER is a Mexican-Canadian electronic artist who works with ideas from architecture, science, technology, and performance. He studied physical chemistry and has a particular expertise in creating site-specific new media installations which poetically connect and explore the digital and physical realms of existence.

GOSHKA MACUGA is a Polish-born artist living in London, UK, who works across multiple mediums such as woven tapestry, sculpture, collage, and performance. She often takes on the role of archivist and curator in her practice, and her work examines the structures of power and control. She was a Visiting Artist at CERN in 2012.

GAVIN PARKINSON is a senior lecturer at The Courtauld Institute of Art, London, UK. Gavin's teaching and writing are concerned with the European art and visual culture of the nineteenth and twentieth centuries with a special interest in French Surrealism. He recently published a book on the Surrealist reception of late-nineteenth-century art, entitled *Enchanted Ground: André Breton, Modernism, and the Surrealist Appraisal of Fin-de-Siècle Painting*.

SUBODH PATIL is a theoretical physicist at the Niels Bohr Institute, Copenhagen, Denmark, working on cosmology, gravity, and related aspects of string theory beyond the Standard Model and accepted phenomenology. In 2013, he was the inspiration partner to the second Collide International artist-in-residence, Bill Fontana.

HIRANYA PEIRIS is a British astrophysicist at University College London, UK, best known for her work on cosmic microwave background radiation. She was one of twenty-seven scientists who received the Breakthrough Prize in Fundamental Physics in 2018 for their "detailed maps of the early universe."

KATARINA PIERRE is the director of Bildmuseet at Umeå University, Sweden. Under her leadership, Bildmuseet has established itself as one of Sweden's key institutions for contemporary art. Katarina Pierre has been instrumental in the creation and the profile of Bildmuseet and in shaping its ambitious exhibition program.

DAVIDE QUAYOLA is an Italian artist living in London, UK. He uses technology as a lens to explore the tensions and equilibriums between seemingly opposing forces: the real and artificial, figurative and abstract, old and new. His varied practice, including sculpture, video, and performance, all derive from customized computer software.

CARLO ROVELLI is an Italian theoretical physicist and writer based at the Centre de Physique Théorique (CPT) of Aix-Marseille University. His work is mainly in the field of quantum gravity, where he is among the founders of the loop quantum gravity theory. His book *Seven Brief Lessons on Physics* became one of the fastest selling science books ever and has been translated into forty-one languages.

SOLVEIG SETTEMSDAL is a Norwegian-born artist who lives and works in London and Bristol, UK. She is a graduate in sculpture at the Slade School of Fine Art and earned her BA in fine art painting and printmaking at the Glasgow School of Art. She was awarded first place in the Jerwood Drawing Prize in 2016 for her video work *Singularity*.

KATE SHAW is an experimental particle physicist. She works at the University of Sussex, UK, and at the International Centre of Theoretical Physics in Trieste, Italy. Her main research is on the Atlas experiment on the Large Hadron Collider at CERN.

SARAH SZE is an American contemporary artist known for sculpture and installation works that employ everyday objects to create multimedia landscapes. Sze is based in New York and is a professor of visual arts at Columbia University. In 2013, Sze represented the United States at the 55th Venice Biennale.

NICOLA TRISCOTT is the chief executive of FACT (Foundation for Art and Creative Technology), Liverpool, UK. She is the founder and former artistic director / CEO of Arts Catalyst, and she specializes in the intersections between art, science, technology, and society. She is also a principal research fellow at the University of Westminster, UK.

KEITH TYSON is a British artist. His work is concerned with an interest in generative systems, science, nature, and the complexity and interconnectedness of existence. Tyson works in a wide range of media, including painting, drawing, and installation. In 2002, he was the winner of the Turner Prize.

JORINDE VOIGT is a German artist with a background in music and philosophy. Working principally within the medium of drawing, Jorinde Voigt's works have been likened to musical scores, scientific diagrams, or notational thought models. Using a precisely coded system of mark-making, she explores natural and psychological phenomena.

JAMES WELLS is professor of physics at the University of Michigan, Ann Arbor, USA. He works on particle physics theory – studying, in particular, the early universe and also simulating those particle productions in laboratories on earth. He was the inspiration partner for the first Collide at CERN artist-in-residence, Julius von Bismarck.

PETER WEST is a British theoretical physicist at King's College, London, UK, and a fellow of the Royal Society. His work is in supersymmetry and he has constructed theories of supergravity in ten dimensions. These theories combine supersymmetry with general relativity.

CAREY YOUNG is an American visual artist based in the UK who incorporates a variety of media such as video, photography, performative events, and installation into her artworks. Her artistic practice draws from a cross-fertilization of disciplines, including economics, law, politics, science, and communication.

ACKNOWLEDGMENTS

First of all, I would like to thank all the staff of Bildmuseet for making the exhibition *Entangle* happen. My very special thanks to the Director of Bildmuseet, Katarina Pierre, for giving me this opportunity, to the extraordinary curatorial staff Brita, Anders, and Sorcha Mae, and to all the technical crew led by the wonderful Olle and Magnus, and to Sofia for her work on the catalogue. It's been a great experience working with you all at one of the most beautiful galleries in the world.

My deepest thanks will always be with CERN and its physicists. I can't think of anywhere more transformational and life-changing than CERN and its scientists who investigate the world below the surface of appearances. In particular I would like to thank the former Director General, Rolf Dieter Heuer, and the former Head of Communications, James Gillies, for accepting my proposal to come to them in 2009 to investigate creating an arts program and cultural policy for CERN. Thank you for hiring me and trusting that I would make Arts at CERN happen: it was a dream come true. My special thanks to everyone at CERN with whom I worked, in particular Luis Alvarez-Gaume, Julian Calo, Bilge Demirköz, Michael Doser, John Ellis, Fabiola Gianotti, Neal Hartman, Despina Hatzifotiadou, Claudia Marcelloni, Arnaud Marsollier, Kate Shaw, Tara Shears, Ian Shipsey, Christine Sutton, and Renilde Vanden Broeck. These are just a few of the names from the extraordinary community to whom I am forever indebted, together with the over seventy-eight artists I worked with who were brave enough to plunge into the world of particle physics with me – especially Julius von Bismarck, Bill Fontana, Iris van Herpen, Ryoji Ikeda, Gilles Jobin, Goshka Macuga, and Jan Peters. Swiss friends who also supported me at this time deserve a special mention: Sabine Himmelsbach, Pascal Hufschmid, and Silvia Wyder, with special thanks also to Antonella Notari Vischer and Bernard Vischer, whom I stayed with in Italy in summer 2018 at their beautiful biodynamic homestay, Podere Trafonti. It was there that I wrote and put together most of this catalogue.

I would also like to thank the Clore Fellowship, who gave me the opportunity to go to CERN in the first place. Without winning the award, I would have never gone there, so my deep thanks are extended to the former director of the Clore Leadership Programme Sue Hoyle, my mentor Sandy Nairne, and Clore fellows who were with me every step of the way, in particular Paul Cutts, Isabel Mortimer, Joana Seguro, and Kathleen Soriano. I would also like to thank Tom Hutchinson. They have all always been there for me and have always been supportive of my wildest dreams to make things happen.

Dedication

To all artists and physicists who work together in the past, present, and future. You create and open up new universes of the imagination and knowing for us all.

ARIANE KOEK
Curator, June 2019

This book is published in conjunction with the exhibition *Entangle: Physics and the Artistic Imagination* at Bildmuseet, Umeå University, November 16, 2018 – April 7, 2019.

The exhibition *Entangle: Physics and the Artistic Imagination* was produced by Bildmuseet and curated in collaboration with Ariane Koek, founder of the Arts at CERN program at the European Organization for Nuclear Research (CERN).

EXHIBITION

Museum director	*Katarina Pierre*
Guest curator	*Ariane Koek*
Curatorial team	*Katarina Pierre, Brita Täljedal, Sorcha Stott-Strzala*
Technicians	*Olle Ebgen-Ljungblad, Magnus Olofsson*
Conservator	*Linda Sandström*
Head of education	*Lisa Lundström*
Education service	*Karin Jonsson, Denise Mångsén*
Visitor service	*Jonas Holmberg, Emma Mårtensson, Karin Nordbrandt, Pierre Persson, Martina Wolgast, Polly Yassin*
Head of communications	*Helena Vejbrink*
Communication support	*Mattias Andersson*
Administrative coordinator	*Maja Sandow, Marie Kårebrand*
Artists	*Julius von Bismarck, Julian Charrière, Sou Fujimoto, Iris van Herpen, Ryoji Ikeda William Kentridge, Rafael Lozano-Hemmer, Goshka Macuga, Davide Quayola, Solveig Settemsdal, Sarah Sze, Keith Tyson, Jorinde Voigt, Carey Young*

PUBLICATION

Editor	*Ariane Koek*
Managing editor	*Anders Jansson, Sofia Johansson*
Texts by	*Philip Ball, Ariane Koek, Gavin Parkinson, Katarina Pierre, Carlo Rovelli, Nicola Triscott*
Interviews with	*Julius von Bismarck, Bilge Demirköz, Michael Doser, Sou Fujimoto, Peter Galison, Iris van Herpen, Goshka Macuga, Subodh Patil, Davide Quayola, Solveig Settemsdal, Kate Shaw, James Wells*
Interviewer	*Ariane Koek*
Photos	*Mikael Lundgren except on pages 96–97 courtesy of Julius von Bismarck; cover image and page 106 courtesy of Solveig Settemsdal; and 128–129 courtesy of Keith Tyson*
Graphic design	*Maria Persson / Plokk*
Project management	*Claire Cichy, Hatje Cantz*
Copyediting	*Dawn Michelle d'Atri*
Production	*Stefanie Kruszyk, Moana Müller, Hatje Cantz*

The exhibition and the catalogue
Entangle: Physics and the Artistic Imagination
were produced by Bildmuseet with financial support
from MTAB and the US Embassy in Sweden

Published by
Hatje Cantz Verlag GmbH
Mommsenstraße 27
10629 Berlin
www.hatjecantz.de
A Ganske Publishing Group Company

ISBN 978-3-7757-4508-6

Printed in Italy

MIX
Paper from responsible sources
FSC® C015829
FSC
www.fsc.org